For earliest STRAWBERRIES and increased profits

its ACCESS Ten by Four

3 weeks earlier than Polythene tunnels
10 days earlier than cloches

A field of Strawberries under ACCESS

ACCESS FRAMES. Light, rigid, galvanised units, 10ft. x 4ft. x 1½ft. high, designed to join together for row-cropping. All glass slides in channels. Frames stackable when not in use. Designed for lifting with glass on.

SELECTED BY THE COUNCIL OF INDUSTRIAL DESIGN

ACCESS TUNNELS and POLYTHENE TUNNELS GO HAND IN HAND

The table shows the good prices from polythene tunnels, and the exceptionally high prices from ACCESS frames. Polythene strawberries start picking as ACCESS strawberries finish; this allows the use of a smaller labour force over a longer period.

Second cropping of Redgauntlet:
ACCESS frames are essential to get the mid-August high prices. Cropping continues into October.

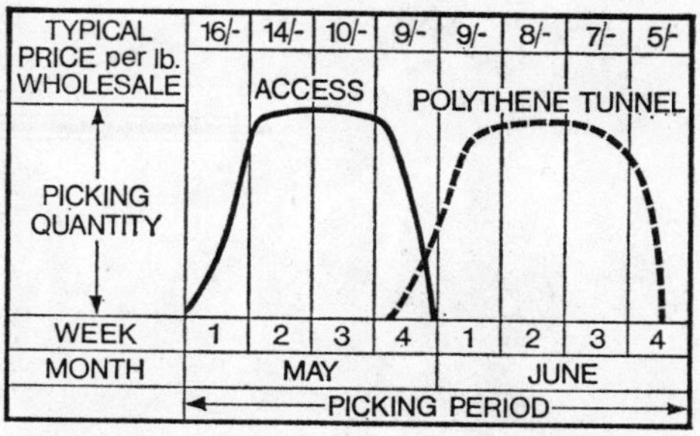

WRITE FOR illustrated brochure and "A profit plan for Strawberry Growers"

ACCESS FRAMES
DEPT SB. CRICK. Nr RUGBY

MINISTRY OF
AGRICULTURE, FISHERIES AND FOOD

Strawberries

Bulletin 95

LONDON
HER MAJESTY'S STATIONERY OFFICE
1970

First published 1937

Eighth edition 1970

The reference to proprietary products in this bulletin should not be construed as an official endorsement of those products, nor is any criticism implied of similar products which are not mentioned.

The Ministry does not accept responsibility for any of the private or trade advertisements included in this publication.

SBN 11 240395 6

Contents

Foreword

THIS, the 8th Edition of this Bulletin has been completely revised and is an up-to-date account of the cultivation of strawberries. All the latest developments in the growing of this crop have been incorporated and it has been prepared under the sponsorship of the Fruit Group of the National Agricultural Advisory Service. The Working Party responsible for this revision has been under the chairmanship of Miss Hilary M. Hughes as the main author with assistance from Mr. J. B. Duggan and Mr. M. G. Banwell. Thanks are due to many growers for information and for photographs. The section on Disease Control has been written by Mr. P. Wiggell and that on Pest Control by Mr. H. J. Gould.

F. W. SHEPHERD
Senior Horticultural Adviser
National Agricultural Advisory Service

Ministry of Agriculture, Fisheries and Food
April 1969

Decimal Currency

References to sterling amounts in this bulletin are followed, in brackets, by the exact equivalent in decimal currency.

General Considerations

THE maincrop strawberry, with the introduction of modern varieties and using up-to-date methods of propagation and weed control, is a comparatively easy crop to grow on reasonably fertile soils. Frosts in May can damage the flowers, so a frost-free site is essential, and sheltered sites encourage growth and crop. Soil drainage must be adequate or various root troubles may occur. Although a wide rotation is advisable and will reduce problems from soil-borne pests and diseases and weeds, many strawberries are grown under little or no rotation, to take advantage of favourable sites. As a result serious soil disease problems can occur.

The major problems now are harvesting and marketing. An acre of strawberries yielding three to five tons over a three to four week picking period requires at least 10 full day pickers, plus supervision. The picking is laborious and not all labour is either willing, or able, to do the job. The fruit is extremely perishable and must be kept picked regularly. Wet weather creates serious problems.

Strawberries offer a quicker return on capital outlay than any other fruit since, under special methods of cultivation, a crop can be picked as early as the first summer after planting. The fruit is the first of the season's home-grown supplies to reach the markets and, while the principal demand is for dessert, it is also used for jam-making, canning and quick-freezing.

The total acreage changes frequently over the whole country. Since statistics were first collected, the acreage has fluctuated from a peak of 26,700 acres in 1935 to the low level of 9,100 acres reached under war-time restrictions in 1943. The area rose to 21,000 acres in 1950, but had declined to 15,200 acres in 1969. On the other hand more careful growing and the provision of improved stocks, following many years of continuous research, have led to higher yields per acre, with the result that total production has increased. Over the 10 years 1957 to 1966 the average annual acreage was 15,500 and the average annual production was 32,000 tons.

The main growing region is in eastern England followed by the south-east and the west midlands and with smaller acreages elsewhere. This distribution has not altered much over many years. The acreage of major producing counties in 1968 was:

Norfolk	3,420
Kent	3,360
Cambridge and Isle of Ely	1,720
Lincoln (Holland)	1,300
Hampshire	800
Essex	790

followed by Worcester, Hereford, Somerset, Suffolk, Devon, Cornwall and Lincolnshire (Lindsey). Norfolk, Cambridge and Lincolnshire are mainly concerned with production for processing and are adjacent to the factories, which also process other horticultural produce grown in these areas. However, a large proportion of the Kent crop is sold for fresh fruit because of proximity to the London markets and, depending on the prices and markets, a proportion of the crop in the other three major producing counties is sold for fresh fruit. Production in the other areas is mainly for fresh fruit, either sold locally, or through the wholesale commission markets.

Large areas of strawberries on any one holding are exceptional; this crop tends to be produced by many growers from quite small units in comparison with other horticultural crops. Statistics show that in 1966 there was only one holding with more than 100 acres and only 200 units of more than 10 acres. These larger units comprised 4,300 out of the total 15,800 acres grown in 1966. But the remaining 11,500 acres were grown on no fewer than 8,500 holdings and most of these were below three acres in extent. This fragmentation indicates small scale production, often on a family basis, and also the difficulty of coping with the picking of this crop on any large scale.

In the major strawberry areas there is competition for suitable labour. Elsewhere female casual labour may be more readily available, but would need training and is unlikely to pick so quickly. Thus choice of a suitable site and provision of adequate labour are two essentials for successful growing, plus easy access to markets.

There has been very little development in farm pre-cooling of fruit or in refrigerated transport and so fruit picked one day must be sold, and generally consumed, on the next. Strawberries cannot satisfactorily be reconsigned from one market wholesaler to another, nor do they store successfully. Because consignments from individual growers are usually small, many wholesalers arrange to collect fruit in the main growing areas. Some growers have also arranged co-operative transport by road, or occasionally by rail. Thus in the non-traditional strawberry areas the advantage of labour availability may be offset by marketing problems. However, isolated producers of strawberries often sell all their produce direct to local shops and hotels and these outlets can be most satisfactory.

The production of early strawberries, grown under protection and ripe in May and early June, is concentrated in the counties of the west and south where warm autumn and spring weather aid early growth, flowering and ripening. Various methods for protecting and forwarding the crop are used. There is much site variation and early strawberries can be produced on warm sheltered sites elsewhere. Early strawberries meet a good demand and price, and production could be increased. It is doubtful whether the production of forced strawberries from heated glasshouses is economic. Market prices in April are affected by imports of strawberries from the U.S.A., Europe and elsewhere, which are now readily available by air freight, and these prices may not always give a sufficient return for the use of heated glass to force strawberries, in comparison with other crops.

Late strawberries from colder areas in the northern counties, or from elevated, or north facing, fields, often meet a ready sale in August, when the main production period of late June and July is over. A recent development has been the production of autumn strawberries, from varieties that produce second crops, or from perpetual varieties, and demand is satisfactory at this time.

Imported fresh strawberries from Commonwealth countries are free from all duties and tariffs, but at present there is a small tariff (see Appendix) on strawberries from elsewhere. There is also a tariff on imported, frozen strawberries and pulp.

Growth of the Strawberry

THE cultivated strawberry belongs to the botanical family (*genus*) *Fragaria* in which there are about 12 species, some native to Europe and Asia, others to North and South America. The native wild strawberry, found in England along hedge banks and woods, is *Fragaria vesca* the wood or alpine strawberry, a small plant with tiny leaves and flowers and very small, dark scarlet, pippy fruit. This species was cultivated in gardens for many years, together with the 'hautbois' *Fragaria elatior*, whose fruits are very little larger. The hautbois strawberry, unlike *F. vesca*, has separate male and female plants (dioecious), with the stamens produced only on the flowers of one plant and the berry and seed bearing part on separate plants and these needed interplanting to ensure good crops.

During the early eighteenth century new species of strawberry plants were brought to Europe from the new lands of North and South America. In eastern America, *Fragaria virginiana* the hardy meadow strawberry grows abundantly, and along the western coasts of North and South America the less hardy, large fruited beach strawberry, *Fragaria chiloensis* is found. It appears that all modern hybrid, large fruited strawberries have originated from the genetical admixture arising from the crosses between these two American species. Some of the original crosses would have been by chance, but English and French nurserymen at the beginning of the nineteenth century were breeding strawberries to obtain better fruit size and soon achieved varieties such as Keen's Seedling and Black Prince which, although now no longer grown, apparently did not differ much in fruit size or flavour and plant appearance from present day varieties.

It is because of the hybrid (heterozygous) nature of the modern strawberry that such a wide range of flower, fruit and plant characteristics are seen. This has enabled local varieties to be evolved, suitable for a wide range of conditions, throughout the world. During this century, plant breeding became a more exact science and the raising of new strawberry varieties has been carried out chiefly at research institutes using plants obtained from other breeders, particularly from America.

THE PLANT

The strawberry is a low-growing herbaceous plant with a short perennial crown or rootstock on which the leaves are produced close together. Branch crowns and runners are produced in the axils of the leaves, which are normally trifoliate. The fruit trusses are produced terminally on the crown but appear to arise laterally because they are displaced by new crown growth before they emerge. The trusses may each have few, or many, flowers on branching stalks.

The roots, arising from the rootstock, form a fibrous mass consisting of the main roots, on which grow numerous side roots or laterals. As the plant becomes older, main roots are frequently produced rather high on the rootstock and the crowns protrude 2–3 in. out of the soil. In England the main scaffold roots are chiefly produced during the period after fruiting until late autumn soil temperatures become too low for further growth. In the spring the growth of lateral roots is more abundant. The roots of healthy well-grown strawberries in good soil can extend to considerable depth and range and their growth is extremely rapid.

The strawberry plant will continue root and crown growth throughout the year if conditions are suitable, but in most parts of England growth ceases during the autumn and recommences as the air and soil temperatures increase in the spring. The younger leaves remain green in the winter although the older leaves gradually die. In wet or severe winters there is noticeably more winter killing of the foliage than during mild winters. In some parts of the world the winter temperatures are so low that the crowns, unless protected by straw mulch, can be killed, but this rarely occurs in England. Growth of the strawberry plant is much earlier in the south-west and south of England where mild, late autumns and warm, early springs ensure a much longer period of growth than in areas with longer winters.

The strawberry plant grows rapidly in the spring, unfolding new leaves and producing flower trusses. Out in the field the plants are in flower during May and June depending on variety and season but earlier if protected by cloches or tunnels. The plants are in flower for at least four weeks. This long flowering period is due to the fact that the trusses emerge in sequence on each crown and the flowers open in sequence on each truss. Older plants with many crowns have a longer flowering period than plants (runners) formed the previous year. These one-year-old plants usually flower before the older plants and have only a few trusses with less flowers on each truss, and thus produce earlier fruit and generally bigger berries. Older plants produce heavier crops per plant, but of smaller berries, because of the competition for plant nutrients between the numerous berries on the one plant.

Each flower of the strawberry normally has five green sepals and five white petals enclosing the stamens and carpels. However, flowers with numerous sepals and petals are frequently observed.

THE FRUIT

The fruit, or strawberry itself, is not a true fruit in the botanical sense, since it consists of the swollen receptacle on the surface of which are embedded the numerous achenes or seeds.

Self pollination is easily effected in the strawberry because the stamens carrying the pollen sacs are arranged around the stigmas of the carpels within the flower. Although insects and bees are seen on strawberries it is probable that wind movement, or moisture, may be more important in effecting pollination. In some varieties, the earliest flowers to open are often poorly supplied with stamens and the pollen grains may also be non-viable, and this may account for some of the malformation of the earliest fruits, particularly under protected cultivation.

To produce a well shaped berry it is necessary for a high proportion of carpels to be pollinated and fertilized; if some are not fertilized there is no hormone stimulation from the seed to the adjacent portion of the receptacle which, consequently, does not grow, and a malformed berry results. Strawberry varieties vary considerably in the average number of seeds per berry, in their size and colour, and whether they are embedded, or stand out on the surface of the fruit.

There are also big variations in the size of the calyx (or cap) and varieties with large leafy sepals are undesirable as they detract from the bright colour of the fruit itself when picked. Each individual flower and berry is attached to the main flower truss by a thinner stalk. For ease of picking it is desirable that this stalk should be readily severed and some varieties break off much more easily than others. There are also marked differences between the attachment of the berry to the calyx; some varieties pull away easily, others with difficulty, and in others the plug may pull out leaving the fruit rather damaged. This latter point is an important factor in varieties that are to be used for processing where the green calyx has to be removed.

After the flower is fertilized it is generally about four to six weeks before the berry is ripe. Early varieties have a shorter maturation period and most early varieties tend to be lighter in crop than mid-season varieties.

NATURAL VEGETATIVE PROPAGATION

The strawberry increases by the production of runner plants on stolons, as well as through seed. Seed is not a suitable method of propagation for cultivated varieties as they do not come true to type. The stolon, a creeping stalk, is produced in the leaf axil and grows some way out from the parent plant during the summer. At the second node a runner plant is formed and a new stolon arises on the runner plant to continue the runner train, which often branches. The runner plant produces roots, at first a few anchoring roots and then a more extensive fibrous root system. Thus the strawberry naturally propagates itself by this vegetative method of runner production (Plate I, bottom). Once the runner plant is big enough the stolon may be severed from the mother plant and the runner plant removed and planted elsewhere. Many stolons may be formed on a mother plant and, as many runner plants can arise from one stolon train, which branches to form numerous runners and stolons, a carpet of several hundred of these daughter plants may be formed in one season.

Because most varieties of strawberries runner freely under suitable conditions of day-length and temperature, each variety can be readily propagated and its characteristics maintained. All the plants of any one variety have thus been derived from the single, original, new seedling by this vegetative propagation; all are exactly the same as the original plant and are collectively called a clone.

Under cultivated conditions in fruiting plantations the runners produced by the original plants may be entirely removed by cultural treatments and the original plants grown on for two, three or four years as spaced plants. Alternatively all, or preferably some, of the runner plants are allowed to root along the rows, excess runner plants being removed. The rows thus consist of rooted runners and the crowns of the original plants forming what is known as a matted row. This gives a high plant population and high yields under suitable soil conditions, but can result in overcrowding of the crowns, disease and small fruit. Sometimes the original plant crowns die, or are crowded, out.

Because of this natural vegetative propagation a strawberry variety is potentially immortal, but most commercial plantations have to be renewed after a few years because of weeds, or overcrowded rows, or because of virus or other disease problems. Virus diseases (page 68) have been responsible for the disappearance of many of the older strawberry varieties.

RESPONSE OF THE STRAWBERRY TO TEMPERATURE AND DAY-LENGTH

Research work has shown that the temperature and day-length (photoperiodism) have a considerable effect on the growth of the strawberry, probably through the control of the production of plant hormones. The varieties grown in England tend to produce runners during the lengthing days of May until August, while flower trusses are initiated during August until November and these bloom in the following year. With some, but not all varieties, removal of the leaves of the plant after fruiting will increase the number of flower trusses and generally the crop the following year, but not in all circumstances. However, defoliation delayed until late August will check the regrowth of the plant after fruiting and result in a crop reduction. Strawberry plants on very fertile soil well supplied with water may well remain too vegetative and, particularly in cool autumns in northern districts of Scotland, may not make sufficient flower trusses; plants under these conditions should particularly benefit from defoliation.

Because of the response of varieties to climatic conditions, particularly of day-length and temperature, it is not uncommon to find varieties behaving very differently in different regions. Northern European varieties lose vigour, and may fail to runner, and flower profusely but crop poorly when grown in warmer parts of the world nearer the equator, under short-day conditions and where they are never exposed to low chilling temperatures.

Although most present strawberry varieties grown in England produce one crop only per year, this is not so with all varieties. The capacity for flower initiation in the spring, to produce a crop after the normal harvest period, and in the summer to produce a crop in the autumn, is quite marked with the well known variety Redgauntlet, particularly when protected under warmer conditions; in Australia and New Zealand this Scottish variety will crop in a series of peak fruit flushes over three to six months. Varieties grown in California behave in a similar fashion, giving cyclic flushes of fruit over a long period; they would give only single crops in Britain.

The so-called perpetual, or everbearing, strawberry varieties initiate and produce flowers throughout the growing season and often produce flowers on the newly formed runner plants even before these are properly rooted. Most of the perpetual varieties have the disadvantage that at any one time there is no large quantity of fruit to be picked and, although they may flower and fruit from late June until the November frosts, the total yield in Britain may not be greater than from a single-crop variety. However, with the mixed genetical make-up of the modern strawberry, it may be possible to breed varieties to extend the flowering and fruiting season beyond the present normal period of mid-June to the end of July. In certain areas Redgauntlet is being grown to give a worthwhile second, or autumn, crop (see section on Late and Perpetual Varieties).

The great adaptability of the strawberry, the different methods of controlling plant density, growth and crop, the variation between varieties and the constant battle to maintain pest and disease free plants, provide great interest and sometimes reward for the grower prepared to study this interesting crop.

Propagation and need for Healthy Plants

THE strawberry is propagated by planting runner plants, which are produced on stolons from established plants during the summer (page 5). This vegetative propagation perpetuates all the characteristics of the parent.

Many experiments have shown that virus disease infection reduces vigour and yield to an appreciable extent, even when no obvious symptoms are produced. It is advisable to plant only virus tested, or as healthy material as possible.

Most virus diseases are transmitted from infected to healthy strawberry plants by aphids (greenfly), but some are transmitted by soil-inhabiting eelworms, and leaf hoppers. The aphid-transmitted virus diseases are of most widespread importance. Virus diseases infect the sap of the plant, and runners from infected parent plants, although often apparently healthy, will be infected. There is no cure for virus infection. Thus the provision of virus-free runners is of prime importance to the strawberry industry. The availability of virus-tested plants, through the certification and nuclear schemes described below, is one of the reasons for the improved vigour and yield obtained from present day varieties.

TESTED CLONAL STOCKS

It is very important to propagate from clonal stocks of which the history, health standard and cropping capacity are known. (A clone is a race of plants propagated vegetatively from a single plant.) The first testing and selection of strawberries was begun at East Malling Research Station in 1931, and since then tested clones of all the more important varieties have been regularly produced there. When necessary, specially treated and retested clonal stock of existing varieties is produced from time to time, and virus tested clones of new varieties are made available to the Nuclear Stock Association as required for propagation and sale.

Certification Schemes for Runner Plants

In order that growers may buy runners of a known standard of health, the Ministry of Agriculture has for many years arranged for inspection of plants and the award of certificates to those plants attaining the required standards.*

* The conditions relating to certification and the varieties and stocks available under the schemes are liable to alteration from time to time. Particulars of current conditions and an annual register of growers of certified strawberry plants are available free from the Ministry of Agriculture, Fisheries and Food, (Plant Health Branch), Great Westminster House, Horseferry Road, London, S.W.1. The Register is issued as soon as possible after the final inspections have been carried out, usually in November.

This scheme has raised the general vigour and health standard and it is obviously to the growers' advantage to use only certified plants. A copy of the certificate, or the number, should be obtained by the purchaser of plants.

There are two types of certificate, 'SS' or 'A', normally available to the grower. Plants of 'A' certificate are quite suitable for planting for fruiting, but should not be used if 'SS' certified runners are obtainable at the right time. 'SS' plants, because of the special conditions of raising, afford the grower extremely healthy material. Any grower wishing to grow his own early runners would be well advised to use 'SS' plants each year from which to propagate his own runners. Runners can be taken from one-year-old fruiting plants but this tends to exhaust the plant and reduce the crop in the first fruiting year, so a special runner bed, preferably isolated from fruiting beds, for early runners is recommended.

Nuclear Stock Scheme

New clones are raised and existing clones are distributed through the Nuclear Stock Scheme. Plants of approved clones are released by East Malling Research Station as mother plants (Nucleus) to the Nuclear Stock Association Ltd., which has its registered office at Agriculture House, 25–31 Knightsbridge, London, S.W.1. This was formed as a growers' co-operative organization during 1953 for the purpose of:

(a) the propagation of any nuclear horticultural stock, including seeds, which may be required by raisers of Special Stock, grown in accordance with standards laid down by the Ministry of Agriculture, Fisheries and Food, and

(b) the sale of nuclear horticultural stock to eligible raisers of Special Stock.

The scheme ensures that these stocks are owned by an independent body, and at the same time that close and continuous technical supervision is provided by the Ministry of Agriculture until the stocks are ready for distribution to the Special Stock (SS) growers.

The nuclear stock is grown in strict isolation and under special cultural conditions, usually in gauze houses, structures covered with fine mesh nylon gauze which exclude insects. The Association arranges the propagation of these plants on a contract basis with a few selected growers. The resultant crop of runners is carefully inspected and if found to attain the required standard of health is certified as Foundation Stock.

This Foundation Stock is distributed on contract to selected growers who are prepared to comply with the conditions laid down for propagation. Provided that the runners from these plantings attain the required health standards, they are certified as Elite Stock, which is sold by the Association to runner-raisers for producing 'SS' strawberry plants.

Special Stock (SS) Certificate

The propagation system developed at East Malling Research Station now forms the basis of the requirements for the Special Stock Certificate. The essential features are:

1. runner raising is treated as a special undertaking entirely separate from fruit production. No land scheduled with red core (page 65) can be used for the production of certified runners and runners produced on scheduled land cannot be offered for sale;
2. runner nurseries are established well away from existing strawberry areas (one mile from any plantation not entered, or eligible for, 'SS' certification and 200 yd from any private garden with strawberries);
3. runners are propagated on the isolated block system or isolated strip system, to allow systematic roguing and removal of any diseased plants and runners in blocks during inspections. Different varieties entered must not be grown within 12 ft of each other. The plants must be de-blossomed; and
4. only virus-tested clonal material approved by the Ministry for the purposes of the scheme, and normally only stocks that obtained an Elite Certificate in the previous year, are eligible.

The runner beds are inspected twice during the growing season and at the final inspection the plants must be a hundred per cent true to type and of a high standard of health.

'SS' certificates are not issued for any stock found to be infected with red core disease, or appreciably affected with mildew or strawberry aphids, or with any other pest or disease, or lacking in vigour.

'A' Certificate

Runners intended for the 'A' Certificate are subject to less stringent standards than for the 'SS' one. Plants of approved varieties have to be grown to show the normal vigour and habit of the variety and must be not less than 50 yd from any plant obviously infected by serious pest or disease. Plants entered must not be grown on land on which strawberry plants were grown at any time during the previous year. There are no restrictions on plant spacing; plants may be fruited and the runners allowed to root in the alleys. Adjoining beds of different varieties must be separated by at least 3 yd and the intervening space kept free from runners. Two inspections are made during the growing season. The 'A' Certificate is issued for stocks that have been found to be at least 99·5 per cent true to type and substantially free from visible diseases and pests.

RUNNER PLANT PRODUCTION

Isolation and Choice of Site

Although strawberries will grow well on most soils, heavy and tenacious soils are not very suitable for runner raising. Runners have to be lifted in the autumn and winter and, if much soil clings to the plant, the lifting becomes both difficult and expensive. Where irrigation is available very light sandy soils produce good runners, particularly if bulky manures can be used. Peaty soils are excellent for runner production but weed control is often a problem. Runner beds can be situated on fields liable to spring frosts as damage to the flowers does not matter.

There are very few conditions for 'A' stock plants, and these have already been mentioned. For raising runners for Special Stock certification, isolation is more important and the requirements have been listed above. These isolation requirements are designed to reduce any virus infection by strawberry aphids. The wild strawberry (*Fragaria vesca*), however, does not constitute a danger.

Verticillium wilt is a disease that can be carried on the soil or in infected runner plants. The disease is widespread and erratic in occurrence (page 66). Land known to be infected should not be used for runner production.

Soil-borne virus diseases which are spread by certain eelworms (page 61) have recently been recognized and are quite widespread. If the presence of eelworms is suspected, soil samples should be taken to see if the soil is free from their contamination. Runners should not be grown on infested land.

Any grower who is contemplating raising plants under the 'SS' scheme for the first time should get in touch with his local N.A.A.S. Horticultural Adviser in order that the scheme can be fully explained and advice given on the high standards of cultivation and control of pests and diseases required.

Planting the Runner Bed

There are two systems of plant raising, the isolated block system in units of not more than six plants, or the isolated strip system comprising not more than twelve plants (Plate II). In the isolated block system planting is carried out so that runners can be trained into separate blocks, leaving cultivated alleys between the blocks. If any plant showing pest, disease or virus symptoms is found at any inspection, all the plants in the same block can be removed, without coming into contact with the neighbouring blocks.

The distance required between the plants depends on the variety, the type of soil and growth expected. Spacing varies from 4 × 4 ft (2,720 plants per acre) to 6 × 6 ft (1,210 plants per acre).

The first runners are laid along the rows to allow cultivations to continue, but later runners develop throughout the block and are allowed to spread beyond the parent plants, but are restrained from reaching into the adjoining blocks.

In the isolated strip system, plants are set out in continuous rows down the field 18 to 24 in. apart, in rows 10 to 12 ft apart. After every twelfth plant a minimum gap of 2 ft is left. The plants are trained by switching all runners in one direction, at right angles to the row, and a gap must be kept between each run of 12 plants. One advantage of this method is that mechanical cultivation can be maintained ahead of the runners, providing a weed-free friable soil to encourage the runners to root. If any diseased plant is found on inspection the whole row of 12 plants and runners is removed.

Cultural Operations

The appropriate certified runners for propagation are usually not available for planting until late autumn. However, it is now permissible to plant cold stored runners certified the previous year. Even if the grower is only raising runners for his own use, certified parent plants should always be used.

Fresh dug runners can be planted in early winter or the spring. Cold stored runners must be planted not later than mid-August, preferably in

July, in order to get them well established. The flower trusses produced a few weeks after planting should be removed. Some runners will appear in the autumn and must be switched into position, preferably within the row of parent plants. In the spring all plants may need firming after frost, and all flowers must be removed to encourage the production of plenty of runners.

As the runners develop they are laid into position according to the system adopted. Runners can be anchored with stones, or soil. A loose friable soil well supplied with moisture, or irrigated if necessary, aids rooting. It is important to lay runners in their final positions before rooting starts, as disturbance thereafter results in a poor root system. The process of laying-in must be repeated several times and, as the block becomes filled with runners, later ones can be laid round the edges, provided that the isolation round the blocks is maintained.

Weeds should be controlled on most soils by a combination of cultivations in the early part of the year and chemical herbicides applied during June, July or August (page 39). On highly organic soils where herbicides are not satisfactory, hand weeding must be carefully carried out. Weeds that are allowed to develop will cause weak, drawn-up runners and make inspection difficult. A good cover of runners will help to smother weeds.

Pest and Weed Control

Careful inspections are necessary, particularly for aphids and mildew. All runner beds should be given routine sprays against aphis in the spring and summer (page 58), and if mildew (page 67), red spider mites (page 59) or strawberry mites (page 59), are noticed, further spraying will be necessary.

Lifting and Marketing

Lifting to supply runners for the early districts, and for cloching, begins in August. The main lifting of runners is during late September and October. Lifting continues, when soil conditions allow, through the winter months until March and April when there is demand for runners for spring planting.

In August, runners are forked up and unrooted runners sometimes bedded back if possible. Irrigation may be helpful, both to lift runners under dry conditions and to re-root runners that have been disturbed.

Strict grading and rapid delivery for planting is necessary. Early lifted runners must be kept well covered and damped down to avoid drying out, and packed for despatch in moist moss, woodwool or peat.

Plants can be lifted in three ways: hand forking and bundling in the field; mechanical lifting with hand bundling in the field; or mechanical lifting and placing runners into trays and bundling in the packing shed.

Hand lifting, by casual labour, under piece-rate working is common practice. This rate includes trimming, grading and bundling into 25's using rubber bands (size 33).

Mechanical lifting, using potato lifters of the hoover or undercutting types, is being attempted by several plant raisers. Plants are then easily bundled in the field or carted in trays to a packing shed. Where several varieties are grown in a field great care must be taken to avoid mixing of varieties.

The yield of runners under either the isolated block system or the isolated row system varies from 50,000 to over 150,000 per acre, depending very largely on soil, time of lifting, aspect and to a lesser extent on season, variety and planting distances.

Various containers such as bushel boxes, orange boxes or paper potato bags are used for despatching runners. Plants should be delivered as quickly as possible after packing. The roots of early lifted plants must be protected, but runners lifted in the winter are often sent out without this protection. Excessive moisture and lack of ventilation in the package can cause heating. Delivery by direct road transport is ideal; passenger train is satisfactory in many areas, but likely times of arrival of the consignments should be ascertained and the customer notified so that rapid collection can be made. If the runners cannot be planted on arrival they should be carefully unpacked, and the bundles laid-in with soil or moist peat round the roots in a sheltered site. On no account must the plants be left in the packing containers, or allowed to get dry.

The Cold Storage of Runners

Also see Short Term Leaflet 54

Runner plants, if lifted when dormant, can be cold stored until August. These plants can be planted in June and July and provide an alternative to fresh dug runners planted in August for fruiting the next year, or for propagation. A few plant raisers have built stores, usually of about 15 ton capacity, specially for the cold storage of runners. Other raisers have used small stores built for storage of pears. Adequate insulation and refrigeration machinery capable of maintaining the required temperature of −1° C(30°F) without fluctuation is essential (Plate IV, top).

Runners for storage must be lifted when dormant, preferably in December and January, but in many seasons lifting can continue until late February. Plants must have a period of cold weather before storing and, in the extreme south-west the mild winter may result in plants never being really dormant so that satisfactory storage is doubtful. Plants lifted in the late winter after cold weather must be stored rapidly before growth commences. Runners are lifted in the normal way, shaken, or washed free of soil, and 300 to 500 plants are placed in a 2 × 2 ft 150g. polythene bag, which should be closed with an elastic band to prevent drying out. On an average, 500 runners can be put in a bushel box depending on runner size. The boxes can then be placed in the store, or the bags put on shelves, but care must be taken to allow a free movement of air around the bags so that the temperature within them is reduced to −1·7 to −1·1°C (29 to 30°F) within a few days of storage. Store temperatures must be maintained as near to −1·1°C (30°F) as possible, preferably −1·7°C (29°F) rather than −0·6°C (31°F) if there is any fluctuation. Rotting of leaves and crowns is often associated with lack of proper cooling, because of too solid stacking, which results in heating of the runners and subsequent development of rots. Infection of the foliage and crown by *Botrytis*, mainly caused by the unsatisfactory control of temperature, is usually the chief reason for loss of runners in store. Whilst plants can remain in store quite satisfactorily until early September, it is necessary to despatch during June and July for planting for fruiting (page 29) and not later than mid-August for runner production. On removal from store quick delivery is essential and it is best to leave the plants in the sealed bags in which they have been stored and allow them to thaw out during delivery.

Mist Propagation

It is possible to root runners rapidly under a mist propagation unit in a cold greenhouse or in Dutch light frames. The mist unit is automatically controlled so that it sprays water whenever the atmosphere becomes dry and thus prevents the tiny, unrooted runner tip from drying out and dying before it has made roots. The unrooted runners with two leaves are cut off from the stolon with a sharp knife (leaving a small piece of stolon to handle the plant) and are closely placed in a suitable rooting compost, or, more conveniently, direct into a pot or soil block. Rooting normally takes about two weeks and then the plants need to be handled without mist, but with sufficient watering, to harden them off before planting out. Plants rooted in compost are usually transplanted at a wider spacing, before planting out permanently.

This method of propagation is useful where stock is limited, as with new varieties, as it enables many more plants to be obtained from one mother plant than allowing the runners to root while attached to the parent. It may also be a useful method to provide healthy plants in disease-free compost in areas where soil diseases prevent propagation in the field.

It is important to ensure that the plants are sufficiently rooted and able to withstand field conditions before planting out or they may be severely damaged. Some planting machines will handle soil blocks or pots satisfactorily and the plants grow away well, without a transplanting check. When planting by hand it is essential to ensure that the block or pot is in good contact with the soil and irrigation is always desirable.

Mist-propagated plants, because of the extra handling involved, are usually more expensive than field dug runners.

Varieties

Through the regular provision of virus-tested plants and the Ministry's certification schemes, together with the use of efficient aphicides, present day strawberry varieties generally remain in cultivation for many more years than in the past, when virus diseases often infected all stocks. New varieties are now raised mainly at the Scottish Horticultural Research Institute, Long Ashton Research Station and the John Innes Institute at Norwich. New varieties from these Stations, from private individuals and from abroad are generally first tested in preliminary trials at the National Fruit Trials in Kent. Promising varieties are further tested at the N.A.A.S. Experimental Horticulture Stations so that their behaviour can be observed on different soils and under different climatic conditions.

PLANT BREEDERS RIGHTS

The strawberry is now eligible for rights under the Plant Breeder's Rights (Strawberries) Scheme 1966 which permits breeders of new varieties of strawberries to have exclusive rights to propagate, or authorize others to propagate, plants for sale or for the purpose of selling fruit. None of the varieties listed below are included in the Scheme, all having been in commerce before it came into operation on 3rd October, 1966; but introductions since then may be included and any doubts about whether a variety is protected should be raised with The Controller, The Plant Variety Office, Murray House, Vandon Street, London, S.W.1. Similar schemes exist in some European countries and it is hoped ultimately to have an international convention. The aim is to encourage plant breeders by ensuring some financial reward for their successful introductions.

Descriptions of recently introduced varieties, after testing at the National Fruit Trials, are published in the Short Term Leaflet 77 which is brought up to date and reissued as necessary.

CHOICE OF VARIETIES

The most important variety at present is Cambridge Favourite, probably comprising over 60 per cent of the total acreage. This variety gives heavy, regular and reliable crops and is grown both for the fresh and processing markets. The berry size does not become too small during the later picks, particularly if the plants are grown on soil well supplied with moisture. Redgauntlet, the next most important variety, is extensively grown for the dessert markets, particularly in Kent and Hampshire, and in areas where red core disease is troublesome. Although slightly later to ripen than Cambridge Favourite it generally produces plenty of very large fruits which aid picking and are attractive for the dessert market. Both these varieties will stand some delay in picking and remain moderately firm, even if fully ripe. They comprise the bulk of the strawberry acreage; other varieties being grown to a limited extent.

Cambridge Vigour is an earlier variety giving at least two to four picks before the two main varieties and is popular in early areas, particularly the south-west and in Essex, but not in Hampshire where it now succumbs to red core disease. The fruit is well flavoured but rather soft and berry size becomes small during the season. This variety is often grown only for annual crops. If retained for the second and third years the berries are often numerous and very small. The plants may grow too vigorously on good soils; the variety Cambridge Prizewinner may then be preferred if an early variety is required.

Cambridge Prizewinner is more difficult to grow than the three varieties already described; it makes only moderate root and leaf growth, and is generally less vigorous in the first year. Before 1969, Special Stock certified plants were not available, and disease problems were often present in this variety. However, on suitable soils satisfactory yields can be obtained and the early fruit is of excellent quality and size, so that where this variety can be well grown it is popular under protection, or for early crops from sheltered sites. It is a useful early variety for fresh land where Cambridge Vigour might be too vigorous and give too much small fruit.

These four varieties are the most widely grown: other varieties are produced only in small quantities and for special reasons. A new variety of Dutch origin, Gorella, is being quite widely tried; although not a heavy yielder the fruit is large and early. The especially well flavoured variety Cambridge Rival is produced to a very limited extent, generally for local sales. Two recently introduced varieties, Templar and Crusader, may have a special use on fields infected with races of red core disease that prevent the other varieties being used, both in the open and for protected crops. The old established variety Huxley is still grown to a limited extent, particularly in East Anglia, for the processing trade.

Royal Sovereign is considered the best flavoured variety but yields are low and it is now only grown by gardeners. A very large fruited variety Merton Princess succumbs readily to botrytis rots but is produced by a few specialist growers, often under protection. Talisman was once widely grown but yields and fruit size are now disappointing and it is seldom planted.

For early production under cloches or plastic tunnels, Cambridge Favourite and Redgauntlet, although not really early, are often used, together with Cambridge Prizewinner, Cambridge Vigour and Cambridge Rival for the earlier crops. Gorella also appears promising for production under protection, and where fruit rots can be controlled Merton Princess should be tried.

DESCRIPTION OF VARIETIES

Established Varieties

Cambridge Favourite (Also suitable for processing and protection)

The most widely grown variety because of its heavy and reliable cropping capacity. The fruit remains firm for several days and does not soften or become dark in colour, valuable assets if picking is difficult. It travels well. Not susceptible to mildew, but susceptible to *Botrytis* in wet seasons. Not so susceptible to verticillium wilt as some other varieties, except in East Anglia. Not field resistant to many strains of red core. Tolerant of aphis-carried virus diseases but isolated plants frequently infected with green petal virus. Streaky yellows very occasionally seen. Produces runners profusely (Plate I, top). Makes a dense compact plant on all soils, but the fruit is easily seen and picked.

Raised by Mr. D. Boyes at Cambridge.

Plant fairly large, moderately open, vigorous.

Flowers large, mainly protected by the foliage.

Fruit large, maintaining size well over the picking season, rounded, plump; pinkish-scarlet, turning light red, colouring evenly, somewhat dull. Flesh firm, moderately juicy, pale pink to pale red, moderate flavour. Plugs fairly easily.

Season second-early to mid-late.

Redgauntlet (Also suitable for processing and protection)

A heavy cropper of good habit and maintaining good berry size. Plant growth is sometimes weak, particularly under poor soil conditions. The fruit is well exposed, so this variety is easily picked, but the fruit stalk is sometimes tough. Travels well and is of good market appearance. Field resistant to some strains of red core and to verticillium wilt and not susceptible to mildew or *Botrytis*. This variety has the ability to initiate trusses during the summer, particularly if the first crop is protected in the spring, and then flower in August and will crop satisfactorily, generally with some protection, from mid-September until November, depending on season. Occasionally in very warm years this second crop can be picked in the autumn from unprotected plants. Flower initiation and second cropping is more marked in the early areas. This second crop does not appear to reduce the crop the following year from normal, autumn initiated trusses. Less prolific in runner production than Favourite. This variety responds well to warm conditions and sheltered sites.

Raised in 1946 by Mr. R. D. Reid, Scottish Horticultural Research Institute; introduced by the Institute in 1956–57.

Plant medium-large, open, upright and spreading, vigorous, with sparse glossy foliage rather slow growing in the spring.

Flowers medium size, only partly protected by the foliage.

Fruit large to very large, rounded to spherical-wedged, broad, blunt, often uneven, with few seeds, attractive scarlet becoming dark crimson. Flesh firm, fairly juicy, deep scarlet, only fair flavour. Plugs fairly easily.

Season mid to mid-late, but early under cloches.

Cambridge Vigour (Also suitable for protection)

A fairly heavy cropper of good quality, from one-year plants. Field resistant to some strains of red core, but very susceptible to verticillium wilt and mildew. Sensitive to drought. Attractive fruit that does not travel well and is inclined to be soft if at all over-ripe, so needs more frequent picking than Favourite. Older plants produce far too many small sized berries. Produces plenty of runners and may grow too vigorously on good soils.

Raised by Mr. D. Boyes at Cambridge.

Plant large to very large, upright and spreading, very dense.

Flowers numerous, medium-large, well protected by the foliage.

Fruit fairly large but inclined to be very small towards the end of the picking season, sharply conical, regular, sometimes a little wedged, very attractive orange-red to light red, turning scarlet. Flesh firm, juicy, moderately sweet, good flavour. Plugs fairly easily.

Season fruit from maidens ripens early, but fruit from older plants is later, and mid-season.

Cambridge Prizewinner (Also suitable for protection)

Essentially an early variety with very attractive fruit but not a heavy cropper. Susceptible to verticillium wilt and red core, but berries not very susceptible to *Botrytis*. Rather shy in producing runners. Needs good soil to grow and crop satisfactorily. Important to obtain certified runners to avoid disease problems.

Raised by Mr. D. Boyes at Cambridge.

Plant medium size. Growth moderate to weak, according to locality where grown, and compact.

Flowers medium size, somewhat exposed.

Fruit conical, occasionally wedge-shaped, bright red, good colour, large at first but rapidly loses size. Flesh only moderately firm; pale red, sub-acid of moderate flavour, plugs easily.

Season early.

Cambridge Rival (Also suitable for protection)

A moderately good cropper of good distinct flavour, often with large fruits. Susceptible to verticillium wilt, but field resistant to most strains of red core. Least susceptible to *Botrytis* of all varieties. The very dark fruit colour is a marketing disadvantage. Rather shy in producing runners and needs planting early if crop from one-year-old plants is required.

Raised by Mr. D. Boyes at Cambridge.

Plant tall, upright, fairly open, moderately vigorous with characteristic light yellow-green foliage.

Flowers medium size, generally protected by the foliage.

Fruit large and often wedged, conical, shiny, bright crimson, turning darker red when fully ripe. Tip of fruit may not ripen. Flesh firm, red, sweet and rich. Plugs moderately well.

Season second-early.

Huxley (Suitable for processing)

Was the most important processing variety, now replaced by Favourite which crops more heavily, and is easier to pick. Needs plentiful supply of soil moisture for sustained yields.

Raised by Albert E. Etter of California and brought to England in 1912.

Plant large, flat and spreading, dense, vigorous with coarse, glossy leaves.

Flowers large, protected by the foliage.

Fruit medium-large, round, dark red, shiny. Flesh firm, pale red, fairly juicy. Berries remain firm and may be picked less frequently than other varieties. Moderate flavour, sweet. Plugs with some difficulty leaving the rather hard core.

Season mid to late.

Talisman (Also suitable for processing)

At one time widely grown, both for dessert and processing markets but lack of berry size now chief disadvantage. Field resistant to red core and mildew. Runners profusely.

Raised in 1946 by Mr. R. D. Reid, Scottish Horticultural Research Institute; introduced in 1955.

Plant medium size, very slow to develop in the spring, upright, dense, fairly vigorous.

Flowers smallish, usually quite well protected by the foliage.

Fruit conical to wedge shaped, medium at first, becoming smaller as the season advances. Quite attractive bright scarlet, turning dark scarlet when fully ripe. Flesh firm, juicy, scarlet, fine texture, good flavour. Plugging fair.

Season mid to late.

Merton Princess (Suitable for protection)

Outstanding variety for size of fruit, but so susceptible to *Botrytis* and red core that it cannot be generally recommended. Worth consideration by the specialist grower of early strawberries on fresh land, prepared to control fruit rots and pick carefully and will crop very heavily if carefully grown.

A Royal Sovereign seedling raised at the John Innes Institute and introduced 1956.

Plant large and dense.

Flowers medium size and hidden.

Fruit very large, generally pointed, rather soft, salmon-pink and quite characteristic. Flavour fair.

Season mid to late.

Royal Sovereign

A very old variety introduced 1892, and very susceptible to virus so that planting and maintenance of healthy stock is essential. Generally considered one of the best flavoured varieties but only gives light crops. Fruit bright, pinkish-orange in colour. Very susceptible to red core, *Botrytis*, mildew and virus diseases. Now only of interest to gardeners.

Recently Introduced Varieties

Templar

The plants are vigorous, very dense and leafy and the flowers small on short trusses hidden in the foliage which makes picking difficult. Crops less heavily than established varieties and may make too vigorous a plant. However, it is field resistant to most strains of red core and may be suitable for soils where other varieties will not grow satisfactorily. Mid-season in ripening and quite good flavour. Susceptible to mildew.

Raised by Mr. R. D. Reid of the Scottish Horticultural Research Institute and introduced spring 1964.

Crusader

Similar in growth, vigour, crop and red core disease resistance to Templar. Ripens a few days before Templar and individual berries generally larger than Templar and more like Redgauntlet in appearance, but lighter in total crop. Field resistant to most strains of red core but susceptible to verticillium wilt. Only moderate flavour.

Raised by Mr. R. D. Reid and introduced spring 1966.

Gorella (Also suitable for protection)

The very large, early and glossy fruit of this variety has attracted attention, although yield is not so heavy as Favourite or Redgauntlet. The compact sparse foliage and growth looks promising for protected cropping. Foliage susceptible to mildew. Field resistant to some strains of red core but very susceptible to verticillium wilt. Not susceptible to *Botrytis*. Worth trial for the dessert market, both in the open and under protection as an early variety.

Raised by Miss H. Kronenberg at Wageningen, Holland, and under test at the National Fruit Trials from 1962.

Plant moderate growth, sparse, with coarse thick foliage.

Flowers mid-season, medium size, on thick stalks.

Fruit very large, conical, crimson-red, sometimes with a green tip, well exposed beneath the upright foliage. Flesh bright crimson, juicy, average flavour.

Season second early.

Elista

Has cropped more heavily than Cambridge Favourite and while the early fruits are of good size later fruit tends to run small, particularly if the weather is dry. Late mid-season prolonging the picking period at least a week after Favourite. Fruits do not extend beyond leaf canopy but are well exposed. Shows some resistance to both *Botrytis* and mildew. May be of interest for processing because of its deep coloured fruit.

Raised by Miss H. Kronenberg at Wageningen, Holland and under test at the National Fruit Trials since 1966.

Plant moderately vigorous, rather upright, a little spreading in the third year with an open habit.

Truss of medium length partially exposed, flowers medium.

Fruit large to medium, round conical to conical, sometimes wedge shaped. Colour, orange red becoming scarlet, flesh red. Flavour good, a little acid. Calyx medium, not usually clasping.

Senga Sengana

A variety widely grown on the Continent for processing. As the flesh is dark red it requires no colour additive. Has been tried in this country but has not found favour with growers owing to the relatively small sized berries.

VARIETIES FOR PROCESSING

Most varieties are suitable for jam making and supply and price are more important factors than choice of variety. Higher prices may be offered by processors for specific varieties for quick freezing, or canning, to growers contracting all, or part, of their crop. Processors may, in addition to specifying variety, also lay down conditions relating to the spray chemicals that can be applied, to avoid taint problems. Contract growing for processing is generally not so profitable as growing for the dessert market; however, if high yields can be obtained, contract strawberry growing for local factories, which avoid all marketing problems, is well worth consideration. Many growers offer their early picks of larger berries on the open market and the main picks to the factories. The standard of the fruit required for quick freezing, canning and bottling are much higher than those for jam.

Canning, Bottling and Quick Freezing

For canning or bottling strawberries need to be of a medium, uniform size, with a good flavour after heating. It is an advantage if the core remains in the fruit after plugging as this ensures a firm texture and assists in retaining the shape of the berry after processing. The core, however, should be barely distinguishable from the flesh and never tough. The berries should also take up the artificial colouring normally used to give an attractive product.

Though no variety now grown can be considered perfect in all respects for canning, Cambridge Favourite is one of the best, giving a well-coloured, pleasantly flavoured product. Cambridge Vigour is sometimes inclined to be dull in colour but otherwise satisfactory and, like Cambridge Favourite, the berries are easily plugged. Redgauntlet has proved fairly good though the exceptionally large size of some of the berries from spaced plants and the tendency to break down during processing can be drawbacks. However, from matted rows Redgauntlet will produce a more suitable berry. Talisman is of only moderate canning quality since a large cavity often remains after plugging. Huxley is difficult to plug and has a hard core.

For quick or deep freezing, fruit is needed which will retain a good colour and flavour after thawing. Cambridge Vigour and Cambridge Favourite are both satisfactory although the latter, which is the most widely used variety commercially, is really too pale for this purpose. Huxley is better frozen than canned. Cambridge Rival has given very good results when quick frozen but the colour may be a little too deep for some consumers.

Jam Making

Talisman and Redgauntlet have generally proved satisfactory though the calyx is more firmly attached with Redgauntlet. Huxley and Cambridge Favourite are usually less satisfactory. The texture of Cambridge Favourite lacks sufficient firmness to retain the fruit whole after boiling; Huxley often gives brownish coloured jam. The berry of Huxley and Cambridge Vigour tend to be tough even after boiling.

Climate, Site and Soil

CLIMATE

The strawberry, like other fruit crops, is susceptible to spring frosts. The flowers open in sequence on the trusses and the plant has a long flowering period during April and early May, so that a complete kill of all flowers due to frost damage is unlikely. If the temperature falls low enough, the centre of the open flowers containing the incipient fruit is damaged, shows the characteristic black eye and the crop may be seriously reduced. Unopened flowers and those that have set are slightly more resistant to frost than partially and fully opened flowers.

It is important therefore to select sites not subject to spring frosts and particularly to avoid alleys and frost pockets. On still, clear nights radiation frosts can occur. The soil and plants lose heat by radiation into space and the air in contact with them is cooled. This process may continue until the air temperature falls below freezing. If the land is sloping the dense cold air moves slowly downhill collecting in valleys (frost hollows). Frost pockets may be caused by windbreaks such as woods, hedges or walls which prevent the slow flow of colder air away to the lower ground. Thus plants at the bottom of the field may be frosted whilst those higher up the slope may not be damaged. Low areas of ground will collect the dense, cold air, when fields on higher or sloping ground may escape.

Freezing can also occur during windy conditions when the temperature is low and these wind frosts are much more damaging on exposed fields than on sheltered ones. During April and May radiation frosts are more likely to occur than wind frosts.

Large areas of water, which lose heat less rapidly than the soil, can create locally favourable situations and many of the early strawberry areas are situated in sheltered sites within a few miles of the sea.

SITE

Apart from the risk of spring frost the strawberry always grows and crops best under sheltered conditions. Wind can seriously reduce growth. A long growth period with early springs, warm summers and mild autumns is desirable and provided there is no risk of spring frosts, sheltered sites with slight south or south-westerly slopes are ideal. On the other hand, a site sheltered from the wind but exposed to the north may be satisfactorily used to obtain worthwhile crops from a late variety.

The growing season in the south-west and south of England is longer than in the east and north, but within any major area there can be marked growth variation depending on slope and aspect from field to field.

Shelter belts can be planted to provide wind protection to otherwise favourable sites. Retention and encouragement of existing hedges and woods is often worthwhile.

Strawberries are sometimes grown interplanted between other fruit crops. Planting between cane or bush fruit, or intensive systems of apples and pears is unwise and not recommended. Intercrops of strawberries under wider plantings of top fruit are often planted. Difficulties from spraying and other operations are bound to occur and it must be remembered that the permanent crop should not be neglected in favour of the immediate needs of the strawberry crop. Whenever possible interplanting should be avoided.

SOIL

The strawberry will succeed on a surprisingly wide range of soils, from heavy clays to light sands and gravels. On very light soils irrigation may be necessary to establish runners when planted early and to maintain berry size and quality during the picking season. On most loams and clay soils, provided the plants have rooted well and there are no root disease problems, summer moisture reserves are likely to be adequate, except in exceptionally dry season. But for maximum maincrop yields the heavier soils with good reserves of soil moisture are better than lighter soils. The silty loams and sandy clay loams of East Anglia can carry high plant populations without irrigation and in this area matted rows are grown to obtain high yields of medium sized fruit for processing.

Lighter soils tend to produce a smaller plant which often flowers a little earlier. These soils will have slightly higher temperatures than heavier soils in the spring and growth will start a little earlier. This may be quite a valuable asset and, because early strawberries can be a profitable crop and give high returns per acre, expenditure on bulky organic manure to improve such soils for early strawberry production may be well worthwile.

In some areas very stony soils are popular for strawberry growing; the stones act as a mulch to retain moisture, yet heat rapidly to warm the adjacent air. They may also provide such a stony surface that strawing is unnecessary. Herbicides can now reduce the cultivation difficulties associated with very stony soils.

The strawberry will not tolerate impeded drainage or waterlogging and plants will be killed. Well-drained fields are required and it is important to keep ditches cleared and outlets from tile drains should be examined each autumn and cleared as necessary. Compaction of otherwise well drained soil, through misuse of machinery under wet conditions, may cause local waterlogging with resultant root damage. In some areas where there is a high water table during the winter, this may cause the death of the deeper roots of the plants. Even if plants are not killed by waterlogged conditions, a part of the root system may be damaged, resulting in poor growth and crop.

Soil infected with verticillium wilt or red core disease (see section on Pests and Diseases, page 56) which prevent normal rooting, will give a poor, or complete failure of, crop. Certain soil-living eelworms can transmit virus diseases and their presence, if suspected, should be checked by tests before planting. Strawberries are often chosen as the crop to follow cleared woodlands and there is a particular danger from eelworms under these conditions. If no eelworms are present, cleared woodland soils rich in humus are very suitable for strawberries.

On alkaline soils and particularly those derived from chalk, or containing much free chalk, strawberry plant foliage may become chlorotic, caused by lack of iron absorption by the roots. Provided there are sufficient moisture reserves and good depth of rooting, moderate chlorosis does not affect cropping.

On large holdings where a wide rotation of crops is possible, strawberries can follow any suitable crop. Potatoes are often suggested but there is a possibility of the carry over of verticillium wilt from the potatoes. Corn crops or leys, or bulbs, are satisfactory preceding crops, although there is a possibility of infestations of stem and bulb eelworm from oats and bulbs. It is therefore particularly important that strawberries for runner production should not follow oats or infected narcissi bulb crops. Advice on the suitability of sites for strawberry growing can be obtained from the local Horticultural Adviser of the N.A.A.S., who will also arrange for soil testing, if necessary.

Manuring

MANURING

As the strawberry may be cropped for one year only or for up to three years or more, manuring should vary according to the time the plantation is to be retained. Before planting, a sample of the soil from the site should be analysed and an estimate made of the requirement of lime, potash and phosphate. If a dressing of 20–30 tons per acre of farmyard manure is ploughed into the land, and any soil deficiency of potassium or of phosphorus is remedied prior to planting, further fertilizer applications will not usually be needed where cropping is limited to one year. A pre-planting dressing of farmyard manure often provides all the needs of the crop over several years on fertile soils. The availability of nutrients will depend upon the extent of the roots which will be decided by the depth of soil and the availability of water in the soil.

NITROGEN

While nitrogen is necessary for plant growth, the strawberry is able to obtain its needs fairly readily. Nitrogen is available to the plant from natural sources in the soil and from added fertilizers and its availability depends upon adequate water in the soil. Recent experiments have not shown response in cropping to nitrogenous fertilizers. The plants used in these experiments were healthy, virus-tested clones which normally made vigorous growth in their early years and so did not need added nitrogen. Too much nitrogen may be detrimental to strawberries, resulting in excessive vigour and even a reduction in cropping due to the vegetative state limiting the initiation of flowers.

There is no evidence of any advantage from the use of organic forms of nitrogen such as hoof-and-horn or meat-and-bone manure in place of inorganic forms.

PHOSPHORUS

Experiments have shown that under conditions in this country the need for phosphatic manuring is likely to be small except in the year of planting or where there is acute soil deficiency. An application of phosphorus prior to planting is normally adequate for the life of the plantation.

POTASH

The importance of potassium for strawberries has been shown in many experiments. On very potash deficient soils, where farmyard manure has not been applied, the omission of potash has generally had serious effects on vigour, longevity and cropping. Annual applications of potash fertilizer are therefore needed, especially when adequate dressings of farmyard manure have not been used. Some soils, such as silt soils, may have a good reserve of potash and will require less.

Acidity

The strawberry is tolerant of fairly acid soil conditions, but sometimes it fails on a calcareous soil where deficiencies of iron, or occasionally manganese, may be induced. Manganese deficiency occurs especially on soils with a high organic matter content. Liming should therefore be restricted to acid soils, but the advice of the N.A.A.S. should be sought and the soil sampled and analysed before any liming is done.

Magnesium

The magnesium requirement of strawberries is not high, but this deficiency sometimes occurs on poor soils such as those overlying gravels, or where excessive applications of potash fertilizers have caused an imbalance between the levels of magnesium and potash.

MANURING PROGRAMMES

Before Planting

A dressing of bulky organic manure is beneficial. Farmyard manure applied at 20–30 tons per acre is to be preferred. Both deep litter, or broiler house, poultry manure and spent mushroom compost are often available and useful but as they contain more nitrogen, phosphates and potash per ton than does farmyard manure they should not be used at the same rate. Two to three tons per acre of deep litter or broiler house manure are equivalent to 20 tons per acre of farmyard manure. If annual cropping is practised on the same site a regular application of bulky manure will be needed every three to four years.

Deficiencies of phosphate and potash as revealed by soil analysis should be corrected. Sulphate of potash is generally preferred to forms containing chloride, e.g., muriate of potash, which should be applied in autumn only. Lime should be applied only to acid soils and preferably for the crop that precedes the strawberries.

No other fertilizers will be required in the first year except on very shallow soils (often on slopes) or with weak growing varieties, where 20–26 units* per acre of nitrogen may be applied before planting.

Second Season

Apply 50–100* units of potash, according to the soil type and potash status of the soil, after the plantation has been cleared up after picking, which is the easiest time to apply fertilizers, or at any time up to January or February.

No nitrogen need be applied unless the soil is shallow or the plants weak, when 20–65 units* of nitrogen may be applied in January to February.

---

* Unit or plant food unit = 1 per cent of 1 cwt of the particular plant nutrient concerned. Thus sulphate of ammonia (21 per cent nitrogen) contains 21 units of nitrogen per cwt of fertilizer.

Three-Year Plants

After cleaning up the beds apply potash again at 50–100 units* of potash as before. No nitrogen need be applied unless the soil is shallow or the plants weak, when 20–65 units* of nitrogen may be applied in January to February.

When applying fertilizer to strawberries the material should be applied between the rows with as little as possible falling on the foliage.

Runner Beds

It is very important that the runner should be produced in soil which is well supplied with nitrogen, phosphate and potash. The soil should also be in good heart and supplied with organic matter to help in maintaining soil conditions suitable for rooting and growing the runners. Unless the soil is amply supplied with organic matter, for example a peaty soil, or is retentive of moisture, or irrigation is available, a dressing of farmyard manure at 20–25 tons per acre should be applied and ploughed in before the final preparations are made for planting in the autumn. Where adequate irrigation is available the provision of organic material is of less importance.

Any serious deficiency of potash and phosphates should be corrected and in addition a dressing of:

Nitrogen at 60 units per acre
Phosphates at 50 units per acre
Potash (sulphate of potash) 100 units per acre

should be applied and worked in before planting. This will not only help rooting but will ensure that the runners will have an adequate supply of these essential elements when lifted for planting out in the field.

Planting Systems

SPACED PLANTS

Strawberries are traditionally grown in the field either as spaced plants, or in matted rows. In the spaced plant method all runners are kept removed during the life of the plantation; the plants are usually spaced 1 to 1½ ft apart in rows 3 ft apart. These distances allow pickers to work between the rows and the plants on good soil generally meet along the row. On really good soil it may be wiser to space the rows wider than 3 ft (Plate III, top).

Spaced plants intended for one crop only are usually planted closer, generally 1 ft apart in the rows, but unless fairly weak growth is expected it is not wise to reduce the distance between the rows much below 3 ft or pickers will damage the fruit.

With the increasing use of efficient herbicides, and hence less need for

* See footnote on page 24.

cultivation, row and plant patterns may be varied. On good soil the above distances do not always leave adequate room for pickers and some fruit damage results, yet wider spacing tends to reduce yields, particularly in the first year. Growers should plan their planting distances to fit, not only the soil and the variety grown, but the machinery and particularly the method to be adopted for picking. The use of a picking tray on legs over the row (page 78) may reduce damage during picking. Another possibility is to vary the distance between rows, allowing a narrower alley between every other row and picking the two rows from the wider alley. There is little to be gained by planting 1 ft apart in the rows for the first crop and spacing out plants to 2 ft for subsequent crops.

The chief advantage of the spaced plant is considered to be production of bigger individual berries and the fruit is easier to see and pick. The chief disadvantage is the cost of runner control.

MATTED ROWS AND BEDS

In the Eastern Counties, almost all the production of fruit is from matted rows (Plate III, bottom) and this method is used by a few growers elsewhere. In this system runners are allowed to develop and root along the row. It is usual to allow all runners to root, but some growers try and limit the number of plants by light cross harrowing over the rows and removal of the unrooted runners. The original rows are planted about 3 ft apart, with 18 in. between the plants. As the runners develop during July and August in the first year they are brushed into the row, usually by women using short sticks. These are allowed to root, making a mat of runners about 15 to 18 in. wide. Any extra runners are cut off in the autumn by rotary cultivators, or destroyed by directed sprays of contact herbicides, before the application of residual herbicides, leaving an alley of 12 to 15 in. In the matted rows, as there are more plants per acre, and a proportion of these each year are one-year-old plants, the system can, under suitable conditions, give a higher crop per acre, over a longer picking period than spaced plants. This system is very suitable for obtaining heavy crops of medium sized fruit for processing. Overcrowding of plants in a matted row may result in severe competition for water in dry seasons, and this system of growing is not suitable on shallow soils or those without good water reserves. One reason for poor yields from matted rows may be that the crowded plants result in a high incidence of botrytis rots.

With the advent of herbicides there has been some interest in matted beds. In this system two rows of plants are allowed to run into each other completely covering the inter-row area, making a bed 4 ft wide. A path of 18 in. is allowed between the matted beds. High yields can be obtained by this method but very careful attention to manuring and disease control is necessary and picking across the bed is difficult.

In the following years cultivations in both matted rows and matted beds consist of a routine mowing of the foliage after harvesting. The debris is then either raked off, or in many cases incorporated in the soil by the inter-row cultivations and these cultivations also remove new runners.

Large number of runners are not normally produced after the first year, but both matted rows and matted beds can become too crowded. Growers should check the number of runners being produced and with this information and experience may need to vary the planting distances for matted rows to fit the expected growth of each variety on each soil type.

Row Lengths

A recent work study indicated that strawberry rows should not be longer than 70 yd. This reduced to a minimum the time pickers spend walking with their fruit to the tally, or collection point. Much fruit is picked by piecework and as pickers are generally in short supply they should be helped to spend the maximum amount of their time in picking. Long fields should be divided by cross alleys every 140 yd, allowing picking from each end. Collection trailers can be stationed in the cross alleys. The position of moveable shelters within the field, both for pickers and fruit stacking and shading is also worth study. Planting is done only once and, with the advent of herbicides, cultivations are less laborious. It may thus be better to plan the rows to help harvesting operations, even if this means shorter tractor runs for other operations.

Row Position

Drainage is important and the rows should not hold up any natural drainage; if plants are on ridges these will have to run with the slope of the field. On slopes where erosion may be a problem the plants can be planted across the slope and will help to stabilize the soil provided the tractor can work that way.

Number of runners required to plant a full acre

Planting distance	Runners required
8 in. × 3 ft	21,780
1 ft × 2½ ft	17,424
1 ft × 3 ft	14,520
1½ ft × 3 ft	9,680
2 ft × 3 ft	7,260

Time of Planting

Runners can be planted at any time from July until the following April and May, provided soil and weather conditions are suitable. If a good first year crop is required early planting is essential, particularly so in less favoured areas. Early plantings may require irrigation, as runners lifted from July to September are actively growing and will not withstand strong sunshine or dry soil conditions. Runners established before the beginning of September will yield well the first year in most areas; those planted during September may give moderate crops, very much depending on the autumn weather after planting, and the locality. Provided there is no check to establishment, the size of runner does not matter for early planting since the soil is warm and root growth rapid, but larger runners will grow and crop better than small runners if planting cannot be done until mid-September or later.

Runners planted in October and November seldom need irrigation and will readily become established. They grow better than those planted in spring, but will not make plants strong enough to give a worthwhile crop the first summer. Mid-winter planting is not recommended; the runners are best left in the runner bed and planted as soon as possible in the spring.

Runners planted in the spring and deblossomed will make strong plants for first cropping as two-year-old plants. After frosty winters runners may be found to have been lifted by the frost and need refirming in the spring, by foot, or using a light roller.

Some growers planting in the late winter or spring space plants at 2 to 3 ft in the row. These plants should be deblossomed the first year, and an adequate plant population is obtained by allowing the new runners to root in the rows. But wide plantings demand few establishment failures or a gappy field will result; the advantages are cheaper establishment costs both in the purchase of runners and in planting them.

ANNUAL CROPPING

Growers who raise their own runners have a big advantage in that they can time the lifting and transplanting to fit the weather conditions. Most growers of annual crops of strawberries have to produce their own runners in order to ensure August planting and worthwhile yields the following year. It is most important that growers raising their own runners should use certified mother plants and take every care to avoid pest and disease problems during propagation and carry out all the precautions described in the section on Propagation and Need for Healthy Plants.

The strawberry flowers earlier and the fruit ripens a few days earlier on one-year-old plants than on older plants, the berries are usually larger, of good quality and easily seen and the plants can be readily sprayed and strawed. On warm sheltered sites, using early ripening varieties such as Cambridge Vigour, Cambridge Prizewinner, Cambridge Rival, possibly the new variety Gorella or even the standard variety Cambridge Favourite, one to three picks will be obtained from one-year-old plants before fruit on older plants is ripe. The price premium obtainable for this earlier fruit, which fills the gap between the end of the protected crop and the main field strawberries, may make it worthwhile to replant every year.

Annual cropping reduces problems from weeds, and pests and diseases, and there is no expense in runner control. However there are heavy establishment costs in producing, or purchasing, early runners and in replanting each year.

In some early areas where red core and/or verticillium wilt are endemic and where strawberries have been grown with little or no rotation for many years, annual replanting is essential since plants do not grow well after the first crop. Planting a strong, well-rooted runner from clean land, during August, or early September, enables the plant to survive these unsatisfactory soil conditions, to make moderate growth and produce sufficient early fruit to warrant the high growing costs. In these areas most of these plants are protected for earlier crops and consequential higher prices (see section on Early Strawberries).

MACHINE HARVESTING

When machines capable of harvesting the crop are developed, and if varieties are produced specifically for mechanical harvesting, row arrangements and plant populations will have to be adjusted to fit these machines. At present prototype machines are being tested.

USE OF COLD STORED RUNNERS

Also see Short Term Leaflet 54

Cold stored runners (page 12) offer an alternative to fresh dug runners which may be both expensive and in short supply in August. Because cold stored runners have to be lifted and placed in store in January definite orders must be placed then with the runner producers.

The cold stored runners produce flowers within a few weeks of planting and they must be deblossomed, unless they have been planted early and made strong growth, or the crop is reduced next year. The plants then have to initiate flower trusses for the crop the following year. Experiments have shown that cold stored runners must be planted before early July in the north and east of England and before late-July in the southern areas, to get a strong growth and yields comparable with fresh dug runners planted in mid-August. After these dates the crop from cold stored runners is much reduced and in some cases the plants will fail to initiate any trusses and be blind.

Cold stored runners planted earlier, in May and June, can be allowed to crop at once, and need not be deblossomed. However, experiments have shown that this late summer yield is always low and picking costs are high and as this fruit clashes with the main production in July it is not worth attempting. These experiments also showed that it was difficult to get ripe fruit in August, when prices were more remunerative, by later planting, without depressing the more worthwhile crop the next spring.

Cold stored runners planted early in May or June and deblossomed will make good growth and produce plenty of runners during August and September which can be brushed in for cropping the next year. As cold stored runners are more expensive than runners direct from the runner bed, there is little to commend this method. It would be more economic to plant fresh dug runners in April, rather than cold stored runners a few weeks later, provided the land was available.

Cold stored runners planted during June to July are more susceptible to damage from herbicides than fresh dug runners planted August and September. Only local experience can show if herbicides can be safely used a few days after planting, or whether a longer period should be allowed.

The chief value of cold stored runners is that they provide a guaranteed supply of certified runners for this early summer planting for annual cropping. The runners thaw out in transport from the store and they must be planted immediately on arrival and never left to sweat and heat in the polythene bags. The cost is usually about £2–£3 per thousand more than the cost of spring delivered runners.

Certified cold stored runners also offer a certain supply of mother plants for early planting for propagation purposes only. The early planting ensures a good supply of early runners the following year.

Cultural Operations

SOIL PREPARATION

THOROUGH soil preparation is important; it is completely uneconomic to plant strawberries on sites infested with perennial weeds such as docks, thistles, bindweed, couch grass and perennial nettles, as it is impossible to eradicate these later without considerable damage to the plants. Ideally, a spring or summer fallow should precede early planting during August to September. If a ley is to be ploughed out for strawberries to be planted at that time, it should first be rotary cultivated or disc harrowed and 40 units/acre of any nitrogenous fertilizer applied to aid rotting, and ploughed in not later than early June. When strawberries are to be planted in the autumn, winter or spring, following a crop or ley, the soil should be ploughed as soon as possible after removal of the crop in summer. If perennial weeds are present, adequate cultivations in a fallow, or the use of suitable herbicides, must eliminate these before planting. Certain complete herbicides require a minimum interval between applications and planting.

As the strawberry roots are planted 5 or 6 in. down and will continue to grow downwards, it is wise to correct any obvious deficiency in soil nutrients, as shown by a soil analysis (see section on Manuring) by ploughing in half the recommended fertilizer dressing for phosphate and potash and applying the rest on top before planting. Neither element is readily leached by rain and the movement of both can be slow in the soil.

If farmyard or other bulky manure is being used this is best worked into the top soil and not ploughed in deeply to form a mat. FYM can be applied after ploughing and the cross-cultivations will work it into the top few inches as the furrows are worked down, or, if the conditions of the previous crop permit, it can be ploughed in with a partially inverted furrow.

After ploughing, the soil should be cultivated to produce a rough tilth and thus to ensure a good germination of the annual weed seedlings. These can then be killed by light cultivations, or alternatively by the use of a desiccant herbicide, such as paraquat, a few days before planting without further cultivations. This method ensures the minimum of weed seedlings during the first few weeks after planting while the strawberry plants are becoming established, and may permit a complete herbicide programme with no further hoeing (see section on Weed Control with Herbicides and Cultivations).

It is particularly important on clay and silt soils not to consolidate the ground too firmly. The roots of the newly planted strawberry need oxygen as well as soil moisture and warmth. Provided there are no problems of drying out, planting in slightly loose soil is preferable to planting in over-consolidated land, particularly if irrigation is to be applied.

It is unwise to prepare too fine a surface tilth before planting, particularly if a system of no cultivations is to be adopted. Frequent cultivations to obtain a tilth will encourage erosion problems on sloping sites and may also cause silting over the crowns of plants set too deeply. A reasonable surface with sufficient depth for working the planter, or to plant by hand, is all that is required.

In some areas where drainage is poor, or red core endemic, strawberries are grown on ridges. The usual practice is to make the ridges with potato ridgers, roll lightly and then plant by hand or machine on the ridges.

MARKING OUT

If a machine planter is used it is only necessary for the driver to mark a sight line across the first row—or line—with poles in the usual manner. The planter has a setting arm that will mark out for subsequent tractor runs across the site.

If planting by hand on a small scale, pegs and lines of string may be used. On a larger scale, one line can be so laid out and a marker used to draw out further lines across the field. The marker is usually 6 ft or more wide with three tines set in the cross-piece at the required row distances, the end marker running in the last row mark. A tractor can similarly be used to mark larger areas, setting pointed tines the required distance apart on the cultivator bar and sighting the first run with poles. The tractor wheels should be set to avoid running on the planting lines and so over-firming the soil.

It is unusual now to plant on the square, as was once done to allow cultivations in both directions. The extra cost of marking plant positions in the row is not recouped by any saving in subsequent hoeing, particularly where herbicides are used. The distance between plants can be gauged by eye, if planting by hand. With some planting machines this can be done automatically according to the adjustment selected in relation to the tractor speed.

PLANTING

Whether planting by hand or machine, it is important to prevent the roots of the runners drying out. This is especially necessary if planting during July to early September, when the runners are actively transpiring, and almost as important in the spring when searing winds can desiccate the fine rootlets. Thus if a gang is planting and some workers are dropping plants ahead of the planters at the correct spacing, they should not be allowed to get too far ahead. Runners should be kept covered with damp sacks, or in their original packing material if suitable. A good damping over-night is sensible and they should be protected from wind at all times.

Runners should be kept protected in trays or sacks out in the field when planting, and on the planting machine only sufficient runners for one journey should be carried to avoid drying-out.

The larger the runner and the bigger the leaf surface, the more it will lose water through transpiration. This suggests that cold stored runners with little or no leaf are not so sensitive to planting conditions—until they start to grow.

Planting should be done only under good conditions. In late summer and early autumn it is unlikely to be too wet, as soil moisture reserves will not have been replenished; the soil is then warm and root growth rapid. The quicker the planting the better, and rather than wait for ideal conditions planting should be done under surface conditions that may not be ideal, without detriment to land or plants, except in the very high rainfall areas. From late October onwards, the soil will be so cold that root growth will be slow and sensible delay to obtain ideal planting conditions is wise.

The soil should be friable so that it settles back around the roots satisfactorily; hence the reason for not over-consolidating the land prior to planting. Planting should not be done under wet or frosty conditions. In the spring the strawberry starts making new root early and planting done as soon as possible, even if not under ideal conditions, is better than waiting and losing valuable weeks of root growth in the permanent quarters.

Planting by Hand

A hand trowel, dibber, spade, or in certain areas a long-handled shovel are used. The most frequent problem is to make a hole of sufficient depth to avoid doubling of the roots. This is the reason for the practice of trimming the roots of very large runners, but as this results in removal of plant material, it must be a checking process, though less damaging than bad planting with doubled roots.

On soils liable to smear, a dibber or spade may produce a most unsatisfactory hole, the sides of which may be impenetrable to the roots and trowel planting is better. But dibber planting in good fertile soil, with plenty of humus in it, is faster than trowel planting.

The runner should be planted so that the crown or rootstock is level with the soil. If planted too deeply the crown may get covered with soil, the leaves then have difficulty in emerging and the runner may die. If planted too high the roots are partially exposed, new roots emerging from the base of the crown cannot root readily, the plant tilts and is readily broken off in subsequent cultivations.

The runner should be placed in the prepared hole at the right height and the soil firmed back with the implement. In loose soil the foot may be used to firm the runner, particularly if the soil is dry, but this is unwise under wet conditions. An experienced planter, if the plants are dropped ahead in position down the row, should plant about 200–300 runners per hour.

Planting by Machine

Several planting machines are available that are suitable for strawberries. The simplest types are those with two coulters which open a furrow and the operator, sitting on the machine, bends down and holds the runner in the open furrow. Two press wheels—partially weighed down by the weight of the operator—press the soil as it falls back around the plant. The depth that the runner is planted and the distance between plants depends entirely on the skill of the operator.

More expensive planting machines use snap, or rubber-coated grip fingers, or spring-loaded discs to hold runners, which are then carried down behind the coulter and released as the press wheels firm the soil around the plant. Again, depth of planting depends not only on the adjustment of the machine, but the accurate placing of the runners within the fingers or discs of the machine.

Provided soil conditions are good, the machine accurately adjusted and the tractor makes steady progress over even ground, machine planting can be better and more uniform than hand planting. Under dry conditions,

Strawberry plants variety Cambridge Favourite showing production of stolons and runners

Strawberry runners as removed from parent plant showing range of size. The very small runners are only suitable for planting early

PLATE I

Propagation of SS strawberry
(*Top*) Isolated strip or row method
(*Bottom*) Isolated block method

Photos: K. E. Cockrill

PLATE II

Established three-year old plants of Cambridge Favourite grown in herbicide treated soil. 1½ x 3 ft.
(*Top*) Spaced plants now meeting in the row
(*Bottom*) Matted row with additional runners fruiting

PLATE III

Inside cold store showing runners in polythene bags in boxes and space for air circulation from cooling unit

Cloched strawberries and use of polythene blind

PLATE IV

Continuous polythene tunnels showing buried ends for anchorage

Tunnels can be readily opened for ventilation or spraying

PLATE V

Applying residual herbicides using fan jets on hand-propelled, powered pump machine

Controlling runners by application of dessicant material through vibrating dribble bar

PLATE VI

Growth stages for timing protective fungicide sprays against grey mould disease
(*Top*) Early flowering. (*Middle*) Full blossom. (*Bottom*) Petal fall

PLATE VII

Photo: Rothamsted Experimental Station

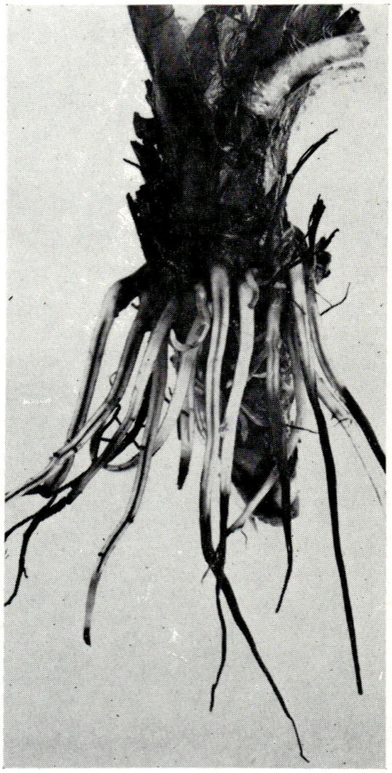

(*Top*) Plant infested with stem and bulb eelworms showing typical leaf distortion and stem thickening and shortening. (*Bottom left*) Strawberry tray carrier in use over rows. The picker is incorrectly holding the berry and not the stalk. (*Bottom right*) Red core disease. Symptoms in roots of infected plant cut to show central discoloration

PLATE VIII

machine planting is generally better than by hand, since moist soil is folded back around the roots.

Plants with very extensive root systems are difficult to plant by machine—as by hand—and the roots may get doubled. Under these conditions root trimming to the length possible to plant is necessary. Not all machines will travel slowly enough for very close plant spacing; the speed at which the operator can work is too slow for the tractor speed. A tractor with a reduction gearbox is generally necessary. The row distances may also be limited by the spacing of the planting units—usually two—and the tractor wheels. Care should be taken not to run the tractor wheel where the next row is to be planted. However, machine planting is so much faster than by hand that it may be preferable to fit the plant rows to the tractor wheels and planter, and if necessary compensate by altering the in-row spacing between plants. Machine planting should not be attempted under wet and sticky conditions when traction is difficult, and it is under these conditions that hand planting may still be necessary.

A planting machine with two planters, supplied with properly graded plants well laid into the trays in front of the planters, should plant about 2,000 plants per hour at $1\frac{1}{2} \times 3$ ft spacing. Where a grower is lifting his own runners, machine planting enables him to lift during the morning and early afternoon and plant in the afternoon, thus ensuring cooler night conditions on the newly transplanted runners. This is particularly valuable during hot weather in August or September.

Cold stored runners are sometimes difficult to plant by machine because there are few or no leaves by which the plant may be gripped. But if care is taken the crown can be held and good results obtained.

DEBLOSSOMING

Deblossoming is the removal of the flower truss; it is done to prevent the plant fruiting, as in special runner beds. Deblossoming should be done when the truss has lengthened, using a knife or scissors. The truss should not be pulled off or the crown may be damaged. If left until the flowers have set and the fruit is swelling, deblossoming will be less beneficial to the plant.

Cold stored runners must be deblossomed in the autumn after planting, unless they have been planted very early and have made strong growth, as the fruit so produced—which is generally worthless—will depress flower initiation and hence the crop the next summer. Late winter, or spring planted fresh dug runners will also benefit by deblossoming in the spring, if labour is available.

STRAWING

Strawing must not be done too early as the risk of frost damage to flowers and fruit is greater once the straw is laid. The right time is just before the developing fruit begins to weigh down the truss to the ground.

The straw should be clean and free from weed seeds. Barley straw is preferred as it is more pliable than wheat straw and can be more easily tucked around the plants, under the fruit trusses. Oat straw may be a source of stem and bulb eelworm, and should not be used. About one ton of straw per acre is usually required; one-year plants that are fruiting can be efficiently strawed with less.

Straw nowadays is usually baled into approximately half cwt bales which can be carted into the field and down the rows on a pallet transporter, or buck rake, on the tractor and can be dropped as required. One operator opens the bale and spreads the straw roughly between the rows. Other workers follow, often on hands and knees, and lay the straw carefully under the fruit trusses, lifting the leaves and trusses with the flat of one hand and tucking the straw in with the other.

On very stony or coarse sandy soils it is not always necessary to straw. Fine sandy silty loams are the most tenacious and more inclined to stick to the berries after rain. When herbicides are used and the soil surface is not disturbed, strawing may not be necessary, with saving in costs.

Some growers have blown chopped straw over the plants as they come into growth in the spring, using forage harvesters and a small amount of hand labour, but at present most strawing is still done by hand and the job is often useful to gather pickers together before picking begins.

It is usual to allow 40 woman-hours per established acre to lay the straw, plus labour for carting.

CONTROL OF RUNNERS

If a matted row is being grown, the runners should be allowed to root along the rows until sufficient crowns are thus obtained. Runners can be brushed into position by workers moving quickly along the rows using short sticks; this must be done before the runners are rooted. Alternatively, light weeding attachments on tractors can be set to pull the runners into place, but this may cause soil disturbance, undesirable where the ground has been treated with herbicides. Excess runners not required in the matted row should be removed and where spaced plants are grown all runners must be removed.

This operation is difficult to mechanize completely. Rotary cultivators are frequently used in the alleys to destroy runners, but runners in the spaced plant rows need brushing out by hand before rooting, or if rooted, need chopping out with hand hoes, or cutting off with knives or hand sickles. Runners should not be pulled off as they may break the branch crowns of the parent plant.

Some chemical control of runners can be achieved with the desiccant paraquat used at the rate of 4 pt per sprayed acre in sufficient volume of water. The finer the mist spray the better the killing effect, but unfortunately the greater the drift and the possible damage to the parent plants. The methods of application are detailed in the section on Weed Control with Herbicides and Cultivations. Adjustable vertical guards to shield the row of plants from the spread of paraquat in the alleys are well worthwhile (Plate VI, bottom).

Paraquat is most effective on dull, cool days; hot sunny days reduce its efficiency. It will also effectively kill annual weeds. Runner control using paraquat works well in summer and early autumn, but rooted runners in winter and spring are more resistant and may need several applications.

There may be some translocation of paraquat from the runner to the parent plant, causing yellowing of the young leaves in the centre. Cambridge Favourite is particularly susceptible, but it does not appear to reduce the crop. Translocation takes place most noticeably in the spring. Direct paraquat damage will show on plants as brown dead spots on the leaves with some chlorosis.

Flame throwers have been used to burn down the alleyways and destroy runners, and machines are being tested. Chemical means of inhibiting the production of runners have not yet shown promise.

CLEANING AFTER HARVESTING

After picking, the fields are often somewhat weedy, particularly if herbicides have not been used and the season has been wet. It is important to clean the beds immediately after harvest since the plant rapidly makes new roots then and is starting to initiate flowers for the next crop. Delay in removal of weed competition can reduce the vigour of the crowns for next year's crop.

With most varieties, prompt removal of the foliage by mowing off all the leaves with a hay cutter bar, rotary cutter or forage harvester, or on a smaller scale using a hand rotary scythe, or shears and sickle, generally improves the number of trusses initiated by the plant and the crop the following year. However, if this mowing is delayed for some weeks after harvest, the crop will be reduced and also, if the mowing is followed by very hot dry weather, the plants may be debilitated.

In general, the hygienic value of rapid removal of the old leaves, particularly on spaced plants, outweight any disadvantage that may occasionally occur. The mowing implement should be set at least 4 in. from the ground. This mowing will also sever some of the runners. The wilted leaves, weeds and straw should then be removed.

Provided there is a good brisk wind going up or down, not across the rows, burning is by far the best method. If the straw has become much entangled with the runners and trodden into the soil, during a wet picking season, it is difficult to burn unless it can first be scratched up by a cultivator, or by hand, using forks. If by hand, it is well worth while to toss some along the row to encourage a good fire along the crowns. Although late July and August are often wet times, there are generally periods of dry weather when burning can succeed. Burning reduces the risk of botrytis disease starting from old plant debris in next year's crop, and leaves an excellent, clean surface, ideal for herbicide applications. It is important to get good burning down the rows, hence the need for wind. In hot dry summers a slow-burning fire may kill out the crowns of the plants; similarly, plants suffering from red core or verticillium wilt may not stand burning. Mechanical flamers can give an excellent burn over fields and are being tested.

Plants grown in matted rows are generally not mown or burnt. It is more usual to use a rotary cultivator with the blades spaced out, to turn in the straw, weeds and runners, leaving about 18-in. strips containing the original plants and the self-set runners.

After the field has been burnt over, or rotary cultivated, the soil should be clean enough for the immediate application of the herbicide simazine. If the soil is very dry, irrigation before the herbicide is applied can help in the regrowth of the plants and also to activate the herbicide. Fertilizers are best applied to fruiting beds after they have been cleaned.

Weed Control with Herbicides and Cultivations

BEFORE the development of herbicides, weed control operations were inseparable from those causing soil disturbance, and the highly beneficial effect of keeping the crop free from weed competition may have implied that cultivations themselves were necessary operations. This is not necessarily so, and in the absence of weeds the strawberry crop grows as well with non-cultivation as with cultivation.

Weed control in strawberries was always an expensive item and weeds often became unmanageable, especially during wet weather and during the long period from strawing until after picking. For these reasons many growers have now adopted systems of non-cultivation using herbicides to control the weeds; others use a combination of some cultivation after picking and in the autumn to turn in the plant debris and get rid of runners, and then use herbicides in late autumn to avoid winter and spring cultivations.

HERBICIDES

If the ground is very uneven after planting, it may be wise to cultivate lightly so that subsequent operations can be easily done, even if no further cultivations are to be done. Soil left in large clods after planting and then treated only with herbicides is too rough for satisfactory strawing down and picking and the weed control may be unsatisfactory.

The chemicals available for weed control are of two types:

residual herbicides applied to the soil; and
contact herbicides applied to the growing weeds.

Herbicides that have been approved under the Agricultural Chemicals Approval Scheme* carry the approval mark on the label (page 74).

Residual Herbicides

The three materials mainly used are simazine, chloroxuron and lenacil, they are relatively insoluble compounds, which when applied as a spray over the soil are absorbed by germinating weed seedlings. They are harmless to strawberry foliage and must be applied to weed-free soil where they will kill germinating weeds for a period of several weeks or months. These materials are tolerated by strawberry plants at the recommended rates and times, but excessive dosage or wrong timing can cause damage and must be avoided. Where damage occurs it is due to the herbicide reaching and being taken up by the strawberry roots, resulting in yellowing, or death of the leaf and a check to growth.

The degree of safety and efficiency with residual herbicides depends upon the rate of application, the soil type, and the age of the strawberry plant. Soils with a high silt, clay, or organic matter content tend to adsorb

* See details on page 74

the herbicides making them less effective and so higher rates of herbicides may be needed on these soils. Conversely, on sandy soils herbicides are very efficient but more likely to cause damage and, therefore, lower rates are appropriate. As these residual herbicides will not kill established weeds they should always be applied to clean soil. If the surface is very uneven and cloddy, weed seedlings may be able to germinate without coming into contact with the herbicide. Maximum effect is obtained when residual herbicides are applied to an even, moist tilth where rain follows shortly afterwards. Failing rain a light irrigation may be needed. Newly planted strawberries are much more susceptible to damage than established plants. If either newly planted runners or established plants require irrigation, this should be done first and the herbicide then applied to the moist soil. This will avoid the risk of washing the herbicide into the soil where it may cause damage.

Simazine

Simazine is a residual herbicide with no contact action but will control *small* weed seedlings. It is very dependent upon soil moisture for its activity. A wide range of annual weeds is controlled at normal rates, but knotgrass, common orache, some veronicas, fumitory and cleavers show resistance. Simazine breaks down slowly in the soil and, if left undisturbed, susceptible weeds will be controlled for about six months.

On established crops, simazine is used after harvest following mowing off or burning, and a repeat application may be made in November or December. Well established young plants may also be safely treated in November or December.

On silty and clay soils, simazine at *low rates* can be safely applied to newly planted runners, but there is always some risk and it is necessary for each grower to determine the safety margin under his own conditions, and at the different planting times. Varieties differ in their susceptibility to damage although this is of less importance than the type of soil and the state of the plant. If experience has shown that damage occurs, this can be avoided by dipping the runner roots in activated charcoal before planting. This technique reduces the risk of damage, since any herbicide reaching the root zone will be held by the charcoal and uptake by the plant is reduced. Grade I steam activated charcoal—not wood charcoal—should be used at the rate of 1 lb per 400 runners.

Charcoal dipping is rather messy and adds to the expense. It is preferable to use safer herbicides (lenacil or chloroxuron) after planting, provided sufficient weed control can be achieved, and to follow with simazine later when the plants are well established.

Simazine is formulated as a 50 per cent wettable powder. It is the cheapest of the three herbicides.

Lenacil

This material is less damaging than simazine and, therefore, particularly valuable for applications after planting, in the spring, and on soils where simazine is damaging to young plants. The range of weed control is almost as good as simazine, but it is particularly useful because it will kill seedlings of knotgrass and common orache which simazine does not control. On the other hand it does not control groundsel well, but gives a better control of annual meadow grass than chloroxuron. It is effective over three to four months.

Lenacil is formulated as an 80 per cent wettable powder and the cost is about the same as for chloroxuron and considerably higher than simazine.

Chloroxuron

Strawberries show considerable tolerance towards this residual herbicide, but several weed species, including annual grasses, knotgrass, common orache, annual sowthistle, red deadnettle, veronicas, fumitory and cleavers are not adequately controlled. It breaks down in the soil more quickly than simazine, generally lasting only eight to twelve weeks.

Chloroxuron may be used in the ways outlined for simazine. It is safer than simazine for spring application, or before cloching. Neither the range of weeds controlled, nor the persistence is as great as with simazine and its cost is higher. Chloroxuron is formulated as a 50 per cent wettable powder.

Contact Herbicides

Contact herbicides will kill existing weeds by absorption through their leaves. Since these materials are entirely non-selective, that is, they destroy any foliage contacted, they must be carefully applied so that only weeds and unwanted strawberry runners between the rows are treated. This technique may be used in conjunction with hand-hoeing in the plant rows, or as an adjunct to other chemical treatments. Contact herbicides may also be useful in dealing with annual seedling weeds before planting. Great care must be taken when applying to avoid drift onto the strawberry plants. Best results are obtained during dull weather. Rain after 4 hr does not matter.

Paraquat and diquat are the most used contact herbicides; they have no residual action and are inactivated soon after reaching the soil. Paraquat is particularly effective against grass species, while diquat is preferable on broad-leafed annuals. In most circumstances, a proprietary mixture containing both materials is used at 2 to 4 pt per acre (½ to 1 lb active ingredient).

Perennial Weeds

Perennial weeds are unaffected by the herbicides described above and should be eliminated before planting. During the preceding summer the weeds may be allowed to develop a dense weed growth before spraying. Many weeds are at their most susceptible stage just prior to flowering and this influences the spraying date. The non-selective herbicides recommended are aminotriazole at 4 lb per acre, or aminotriazole plus 2,4-D. A period of *six weeks* must elapse between spraying and planting and, as these materials are toxic to most crops, the contamination of equipment and possible drift must be carefully avoided. The use of a separate sprayer is recommended.

If perennial weeds do occur in established strawberry fields, it may be best to sacrifice some of the strawberry plants to ensure clean ground when the crop finally is ploughed in. Spot treatment with hormone sprays of MCPA or MCPB in May and June will control bindweed but cause some strawberry plant damage. Chlorthiamid or dichlobenil granules can be used to spot treat couch, thistles and docks in spring or autumn. These latter two materials will kill the strawberry plants amongst the weeds, so care must be taken when applying the granules to avoid unnecessary scatter. Hand application from a small shaker container is the most satisfactory method for spot treatment.

Recommendations for Weed Control in Strawberries
(*See current Short Term Leaflet* 23)

Time	Chemical	Rate of Active ingredient per acre in 50–100 gal	Remarks
1. *New plantings*			
One or two days before planting	Paraquat and diquat	1–2 lb	If necessary to destroy existing seedlings
Shortly after planting	Lenacil or chloroxuron or simazine	2 lb 4–5 lb ½–¾ lb	Simazine only if known to be safe under known soil conditions
followed by			
Late October to early December	Simazine or lenacil	½–¾ lb 2 lb	For runners planted August to early September
March to April (before flowers open). Earlier under protection	Chloroxuron or lenacil	4–5 lb 2 lb	Unnecessary where good weed control achieved by winter application of simazine or spring germinating weeds not expected
2. *Established plants* (*after first summer*). *Spaced or matted rows*			
August to September	Simazine	¾ lb	After burning or mowing, or inter-row paraquat and diquat treatment. If dry, irrigate before application
August to October	Paraquat and diquat	1–2 lb	Runner control between rows; may need repeating
November to December	Simazine	¾ lb	Before weeds germinate
March to April (before flowers open)	Lenacil	2 lb	Unnecessary where good weed control achieved by winter application of simazine
3. *Runner beds*			
End June to July	Simazine or or lenacil	½–¾ lb 2 lb	Apply when runners appearing, to moist clean soil. If dry, irrigate before application.

General Considerations

Having decided on suitable herbicide treatments, it is equally important to ensure even and accurate application. The volume of water applied is not critical, provided it is sufficient to give an even ground cover. Residual herbicides are normally applied in 50–100 gal of water per acre, the higher rates being preferable for small-scale operations, while lower rates may be satisfactory with larger, more accurate sprayers. Manufacturers' instructions should be followed when mixing, and thereafter continuous agitation is essential to avoid the herbicides settling out in the spray tank.

Most residual herbicides are used as overall treatments, and it is important to avoid overlapping of sprayed areas which would result in overdosing and possible crop damage.

Spraying Equipment

Overall Sprays of Residual Herbicides

The type of equipment used depends primarily upon the scale of the operation, but in each case the height of the nozzles, constant spray pressure and uniform speed of travel are important for accurate applications.

Knapsack sprayers are best fitted with a boom and are either carried on the operator's back or on small wheels. There are small machines with horizontal booms and a motorized pump which are hand pushed (Plate VI, top), or spraying attachments fitted on motor cultivators. Tractor-mounted sprayers operate from the power take-off, with the tank and booms mounted on the three-point linkage of the hydraulic system. Normal operating pressure is between 30 and 50 lb per sq. in.

Either ceramic fan nozzles or disc nozzles are suitable, but their pattern of distribution differs. Fan nozzles give a more even spread, but more drift in windy weather than disc nozzles. The correct boom height is less critical for fan nozzles than with cones, and the latter need checking for disc wear, while ceramic fan jets become inaccurate when chipped. (For further guidance consult M.A.F.F. Mechanization Leaflet 2. *Field Crop Sprayers*, price 2*s*., by post 2*s*. 4*d*., obtainable from H.M. Stationery Office, or through any bookseller).

Contact Herbicides

Except where contact herbicides are being used before planting, the following equipment must be used to give a directed application and avoid damage from spray drift, which can give serious leaf damage to young strawberry plants.

1. Dribble bar
2. Flood jet
3. Hooded (guarded) sprays
4. Vibrating nozzles

Dribble bars use the gravity flow of liquid from a tank, giving a small pressure. The suitable equipment ranges from a simple watering can to a shoulder-mounted, hand-held dribble bar, or wheeled machine. Dribble bars to cover several alleyways can be tractor mounted. The dribble technique has the disadvantages that a higher volume of liquid is used, and coverage of small seedlings and runners may be incomplete, with disappointing results. Sprays are more efficient, but if used amongst the crop must be at a low pressure, using either a flood jet which produces little drift, or a well-guarded spray from a fan jet, or the proprietary vibrating nozzle available (Plate VI). For overall treatment of land prior to planting where immediate drift is not important, a fine spray will give the best kill of weed seedlings. Treatment with these contact herbicides must be carried out in calm weather, preferably in late afternoon or evening when several hours without rain can be expected.

Calibration of Equipment

Before applying residual herbicides a knowledge of the sprayer's output is essential, and trial run is necessary to check application rates.

For knapsack sprayers and small-scale equipment, a spraying rate of 50 gal per acre is suggested. This is approximately equivalent to 1 gal on 100 sq. yd. If batches of 3 gal of spray are prepared, this will treat 300 sq. yd (1/16-acre) and the following formula is convenient:

1 lb herbicide per acre = 1 oz on 300 sq. yd
1 pt herbicide per acre = $1\frac{1}{4}$ fl. oz on 300 sq. yd

If, with the equipment available, it is not possible to cover 100 sq. yd evenly with 1 gal of spray, then increase the rate of water to 100 gal per acre, that is, 1 gal on 50 sq. yd or 6 gal on a 300 sq. yd plot.

To avoid the need for frequent measuring and mixing, a larger volume of spray than is needed for one knapsack or tankful can be prepared. Sufficient herbicide for 1 acre may be mixed into 50 gal of water and this applied at the rate of 1 gal on each 100 sq. yd, but it is most important to agitate the spray mix thoroughly before drawing off each tankful.

Tractor-mounted sprayers are provided with a chart giving rates of application per acre at various speeds and pressures, using different sets of nozzles. A low reading speedometer and rev. counter on the tractor are essential. If accurate information on a sprayer is not available, or thought to be unreliable, possibly due to nozzle wear, the following test should be carried out:

While standing on level ground, completely fill the sprayer tank with water. Make a trial spraying of $\frac{1}{4}$ acre while running at a suitable speed and pressure. To cover $\frac{1}{4}$ acre, the distance to travel is $\frac{1{,}210}{\text{length of boom in yd}}$. Using a boom 2 yd in width one needs to travel $\frac{1{,}210}{2} = 605$ yd to be equivalent to $\frac{1}{4}$ acre.

Return to level ground and top up the tank using an accurate measure. The amount of water needed to top up to the original level multiplied by four is the rate per acre. It is wise to check this figure regularly in practice and keep a note of the following for future reference:

(a) Speed m.p.h. (b) Engine speed r.p.m. (c) Gallons per acre.

SPRAYING DO'S AND DONT'S

Do not spray if soil or wind conditions are unsuitable.
While spraying check all nozzles are delivering.
Check output of the first tankfull against a known area, stop if application rate is not correct.
Keep herbicide mixtures well agitated.
Avoid double spraying on headlands.
Mark last row sprayed when leaving site for refill.
Avoid drip from nozzles on to plants; do repairs on headlands.
Empty surplus material on waste ground and wash out sprayer thoroughly.

Cultivations

Provided the soil has been adequately prepared before planting, only annual weeds will need control and light surface cultivations with A or L shaped hoes suffice. Deeper working of the soil with straight tine hoes will be necessary, if the plants are to be ridged in the autumn. This method is still adopted by some growers but has little to commend it as the ridging must damage many of the surface roots. If the practice is to add surface drainage during winter, it is better to plant on ridges and to follow a system of non-cultivation using herbicides and leaving the ridges intact.

Hand hoeing will be necessary for control of weeds in the plant rows. Short-handled draw hoes with sharp blades are used and workers should be trained to draw the soil up to the crown of the plant and thus aid rooting. Hand hoeing, if badly done, can result in damage to the crowns. As weeds are hoed, runners can also be chopped out or pulled into the alleyways.

A woman worker will take about 50 hours to hoe the rows of one acre of strawberries, but this time will vary according to the amount of weeds and runners present. Tractor cultivations of the alleys will take about 6 hours per acre.

The tractor mounted rotary cultivator is often used to turn in runners, weeds and straw after picking. The blades are spaced out, by removing those which run over the plant row, and it is usual to cultivate two alleys at the one time. Although the rotary cultivator will penetrate 6 to 8 in. in depth, it is best to work to the minimum depth sufficient to incorporate the material into the soil, and to increase the speed of the cultivator while travelling slowly, to ensure thorough chopping of the debris and minimum damage to the strawberry roots. Although the rotary cultivator leaves a good looking tilth a hard pan can be formed below this. Soil that has been rotary cultivated is very inclined to erode from slopes under heavy rainfall. Weeds tend to grow abundantly after rotary cultivation.

Early Strawberries

Early strawberries can be produced by protecting plants with continuous glass cloches, or with coverings of polythene. Glass cloches consist of sheets of glass held in place by wires to form a simple tent, or a flat or sloping topped barn cloche over the plants. The cloches are placed end to end to cover the plant row, and the cloche at each end is generally closed with another piece of glass. Polythene film is used to form a continuous cover along a row of plants; the polythene is supported on wire hoops just clear of the plants and is tied down to the hoops and to a peg at each end of the row.

A very small quantity of early strawberries is produced by specialist growers from cold Dutch light glasshouses, and also from Dutch lights on the flat. It is doubtful whether the production during March and April of extra early strawberries from heated glasshouses is economic. The market is very limited and prices have to be high to cover the expenses of production.

The market for early strawberries, ripe during May and early June, before fruit is available from the outdoor fields, is buoyant and there is a good demand for well presented fruit at this time. In general, competitive imported fruit does not look so well as the home produce and also carries a small import duty. Costs of production are high and early strawberries involve much hand labour; this may account for the popularity of protected strawberries on suitable smallholdings, where the occupier can do most of the work himself, employing extra labour only for picking. Under these conditions the grower's own labour spent on detailed jobs to enhance earliness of crop can be well repaid.

The area of strawberries protected with glass cloches is not large, possibly some 80 acres, and is concentrated in those parts of the country that enjoy mild winters and warm springs. Hampshire and the Isle of Wight have by far the largest area of cloches. West Sussex, Kent, Somerset, Devon, Cornwall and East Sussex each have some five to ten acres. In many other counties glass cloches are used to forward strawberries on a small scale.

The use of polythene tunnels (Plate V) is rapidly increasing though there are no statistics available of the acreage covered. This method of protection is far less costly than glass cloches but the crop is not so early. It is generally considered that glass cloches will advance the crop by four weeks, compared with field plantations of similar varieties on similar sites, and polythene tunnels by about a fortnight. Fruit ripened under tunnels is generally cleared before main crop strawberries in the same area are ready to pick.

The prices for early strawberries on the main wholesale markets fall rapidly as supplies increase. Even so, in less favoured areas where a local outlet is assured, it may be profitable to protect strawberries and gain a few days advantage over other local suppliers.

GENERAL CONSIDERATIONS

Choice of Site

Although a light, well-drained soil that is readily warmed by the sun, is a valuable asset, climate is the over-riding consideration affecting earliness. Radiation is mainly dependent on latitude, hence the advantage of the south and south-west over the midlands and north. The presence of the warm seas on the west of England and Wales also confers an advantage over areas in similar latitudes in the east of the country. The value of proximity, but not exposure to the sea, is very marked, not only in direct effects on air temperatures but probably also from increased light.

Records show that the traditional early areas have higher maximum air temperatures than elsewhere. If it is assumed that temperatures above 5·6°C (42°F) are needed for growth, then conditions are much more favourable from January until April in the traditional early areas than in the rest of the country. The south-west has better conditions than the southern areas during January to March, but similar advantages during April. These temperature differences account for the fact that strawberries are ripe in favoured areas in Devon and Cornwall some days ahead of Hampshire and Kent.

Within the main over-riding meteorological differences, aspect can be quite important. Calculations show that ground sloping steeply to the south, as in the Tamar valley, can receive twice as much radiation during the winter months as level ground.

Shelter

Shelter is also an important factor. Tall hedges will cast shadows, particularly during the winter months when the sun is lowest; thus shelter belts running as near as possible north to south, provided they give the necessary windbreak, are to be preferred to those running east to west.

Wind is the biggest disadvantage to early production, since wind movements through cloches and tunnels will remove the air warmed by the radiation from the sun and trapped under the protection. The best increases of air and soil temperatures are obtained on still sunny days.

Because only small areas are covered with cloches or tunnels and because wind can not only reduce temperatures but cause damage to polythene or glass, it may be worthwhile utilizing artificial shelter screens to split up more exposed fields. Fences with 1-in. laths spaced 1 in. apart are ideal windbreaks. Cheaper, but less efficient, screens can be made of doubled wire netting of ½-in. mesh, or of the nylon mesh netting that is now replacing the jute fillis hop netting. All these materials need well braced strong supporting uprights, and are usually about 4 to 6 ft high. Shelter screens should be permeable so that they filter and reduce but do not try to stop the wind. Solid screens, such as walls, thick woods or buildings, may create local wind turbulence and cause damage.

No detailed experiments have been done on the best ways to site the cloched rows for earliness. On level ground the rows are usually run north to south, but on sloping or difficultly shaped fields the lines of cloches or tunnels should be positioned for ease of management.

Soil

Light soils warm more quickly than heavier soils, and well drained soils better than badly, or slow draining soils. The grower has little control over the soil once a site has been chosen, but it is probably more important to opt for good overall shelter and aspect, rather than choose a more exposed site with lighter soil. Assuming no major drainage problem exists, heavy soils can be slightly improved over the years by the incorporation of bulky organic manures, coarse graded peat, or by a system of leys. A system of growing plants on ridges may help drainage and, by increasing the soil surface exposed to the sun, increase soil temperatures.

Frost Damage

Protected strawberries flower some five to seven weeks ahead of the unprotected plants. It is unusual for early strawberries to suffer from spring frosts, because they are grown in areas where there are major climatic advantages and the whole area, including the soil, is warm. Meteorological records show that in these areas the likelihood of spring frosts in March and April is low. However, in less favourable areas, if protected strawberries are well advanced by much sunny weather, radiation frosts could be serious, especially if they follow after a period of dull weather. Glass and polythene give protection only by retaining the heat trapped during the day and, under severe night frost conditions, polythene in particular will give little or no protection.

CLOCHES AND TUNNELS

Planting Distances

The distances will depend mainly on planting time, if one-year plants are to be covered, and to a much lesser extent on the soil and variety. Generally, if cloches are to be used, one-year-old plants are protected; two-year-old plants are rather large and may become leafy and etiolated under cloches. Tunnels are used to cover one-year-old plants, but are also suitable to protect two-year-old plants if these are not too vigorous. In areas where it is difficult to establish a strong one-year-old plant, a good compromise is to plant in April, and allow some runners to root along the row, which is then covered the following year.

To obtain well grown one-year-old plants the grower may:

1. purchase fresh-dug runners in August;
2. raise his own plants for transplanting in August;
3. use cold stored runners planted in July.

It is necessary to grow the right size of plant to fit the cloches or tunnels available and fully to utilize the ground protected. In the traditional early areas, a fresh dug runner planted the last week of August, or during the first week or so in September in the south-west, will make good root and crown growth in the autumn and give a good crop the following spring. Planted at this time, the runners should be about 8 in. apart in single rows, and the plants will be suitably covered by the grower's barn cloche, or a medium, or large, sized tunnel. Planted at this time, provided irrigation is available so that there is no check to growth, the size of runner has been shown to be of no importance, the planting date solely affecting total crop.

After early September, the growth the transplanted runners make before the onset of winter is much reduced and runner size is more important, better growth being made by the larger runners. If planting is delayed until late September, the in-row planting distance can be adjusted to 6 in., or a double staggered row used within the cloche or tunnel.

If tent cloches only are available, August planting may produce too big a plant, and the runners are best planted during September, depending on the locality.

Experience will soon show the best combination of in-row distance and row distance for the protection available. Many growers plant a part of their area in late August or very early September, when runners are rather expensive, and plant the remainder of the area during September when runners are cheaper and more readily available, closing up the plant spacing in the row. The earliest few fruits to ripen are generally obtained from the smaller, late planted runners, but the best total yield from the earlier planted areas.

Runners can be planted by machine. Row distances depend on the soil and method of covering. On sites that have been cropped with strawberries for many years and where red core disease is troublesome and vigorous growth not expected, rows may be 2½ to 3 ft apart. If barn cloches are used, $3\frac{1}{2}$ ft is a usual distance, allowing $1\frac{1}{2}$ ft between the cloches, or a double row of cloches adjoining each other may be planted, giving two rows 2 ft apart and then a gap of $3\frac{1}{2}$ to 4 ft allowing an alleyway to pick from both sides.

Tunnels are generally used over rows 3 ft apart; this is the minimum for a medium or large tunnel and is rather too close for easy working if two-year-old plants, or matted rows, are to be covered with a 2 ft wide tunnel. The single row is favoured by most growers, but in some early areas two rows of runners are planted, generally 12 in. apart, with the plants 6 to 12 in. apart in the row, depending on planting time. These double rows are covered with a wider tunnel or arch spanning at least 2 ft at the base.

As with the field strawberries, it is unwise to have very long rows without cross paths to allow tractors or hand trolleys, since the picked fruit has to be carried long distances. Rows should not be longer than about 200 ft.

If fresh dug runners cannot be obtained, cold stored runners have been shown to be a very suitable alternative (page 29). These runners have to be planted earlier as they immediately produce the spring flowers—which have to be removed—before initiating the flower trusses in the crowns for the protected crop the following spring. Where sufficient land is available, thus making earlier planting no disadvantage, cold stored runners are not only cheaper than fresh dug runners, but also can be reliably booked for delivery from store irrespective of the current season's growth of runners, or conditions for lifting. In the very warm south-west, cold stored runners of most commercial varieties can even safely be planted during early August; elsewhere they should be planted before the end of July, or even much earlier in less favoured districts. If cold stored runners have made very dense and leafy growth, some defoliation at the time of covering in the spring is beneficial.

Types of Cloches

Glass can be used in the form of tent cloches, two sheets of glass, generally 18 × 16, or 20 × 18 in., held together with a wire handle and clip under tension. Tent cloches can also be home made, using a 6 in. length of 2 × 2 in. wood with a groove to take the top edges of glass at the appropriate angle. The glass is held to the wood block by a piece of wire bent round the lower edges of each sheet and fastened to a screw eye on the underside of the block. These home-made cloches are not so rigid as the purchased tension spring type, of which the barn type cloche is most popular, generally consisting of sides 12 × 24 in. and top 24 × 12 in., giving a cloche 2 ft long and 1 ft high and nearly 26 in. wide. The sides can also be 6 or 9 in. high. These cloches are held together by wires under tension, and the wires permit of the removal of one of the top panes of glass to allow access to the plants. Other cloches, some with flat tops, are available, of varying sizes and widths and with different methods of retaining the glass.

Horticultural glass is generally 24 oz per sq. ft and sold in cases containing 200 sq. ft, giving the following number of sheets per size per case:

Glass Sheets (in.)

24 × 6 = 200	24 × 12 = 100	18 × 16 = 100
24 × 9 = 133	20 × 18 = 80	

Use of Glass Cloches

The cloches are generally put over the plants during February. In areas where heavy snowfall or icing may occur, it is wise to wait until after the normal time for expected snowfall to avoid breakages. Cloches should go on level, clod-free soil, an important point to remember when cultivating the surface after planting and prior to herbicide applications. It is unwise to cultivate between the cloches; flying stones can cause damage. Cloches will withstand frost. The use of herbicides is recommended since the cloches can then be put out at any time, without the need to remove them for hoeing.

In early areas, cloching during January to March has little effect on the time of flowering and ripening, but covering after the end of March and in early April will result in later ripening. In early areas, February to March covering is satisfactory, and March covering for later areas. It is not yet known whether winter covering would be beneficial. In very wet years, on soil where drainage is poor and red core disease a problem, early covering after Christmas may help to shed the water into the alleys away from the plants. It is important to place the barn type cloche with all the ventilating lights on the same side so that they can be readily opened.

The ends of the cloches may be blocked with glass, wood or galvanized sheet steel cut to fit the shape of cloche and held in place with a peg, tied back to the handle of the end cloche. Not all growers trouble to seal the ends of cloche rows.

As the plants grow during April and May it will be necessary to space the cloches and remove the ends to avoid excessive heat and etiolation and too dense growth of the plants. During blossoming an even day temperature of about 15·6° to 21·1°C (60° to 70°F) is necessary with plenty of wind movement to help pollinate the flowers and care should be taken to avoid hot, drying conditions. The top ventilating pane of glass can be opened on to the first or second wire notch depending on weather conditions.

Rows of tent cloches can be readily ventilated by removing a cloche and stacking it on its neighbour. The number to remove will depend on the growth of the plants and the weather.

For the aphicide spray in March and for fungicide sprays during blossoming, it will be necessary to move the cloches off the plant, to the alley, or to open the ventilating panes, or to move two cloches at the end, then spray the plants and shift the next two cloches down and so reach all the plants.

When picking commences, many growers remove the cloches and stack them conveniently in the alleys, or they may move them over an adjacent row of plants to get a little benefit; but in areas where birds are troublesome, or if the weather is cool so that protection from the glass is still required, the cloches are left in place and the fruit picked either by moving each cloche down the row, or by removing the top pane. Wire netting can be used over the ends of cloches to keep out birds.

One method to improve access without moving the cloches is to replace the ventilating pane with a polythene blind (Plate IV, bottom). Most blinds are home made; one method consists of two 6 ft lengths of 2 × ½ in. unplaned timber, preferably pressure treated with a preservative. Polythene (gauge 500) is stapled to the top lath so that it fits along the cloche handles of three adjoining cloches. The polythene is cut wide enough to allow the bottom lath just to hang over the side of the cloche. Each blind replaces three ventilating panes which are stacked on the other side of each cloche. Alternatively a long strip of g.500 polythene with the long sides stapled over polypropylene twine can be used and hooked along the cloches. The blinds permit quick access to the plants. At the end of the season the panes are replaced to allow safer handling of the cloches.

Glass cloches are heavy to handle, and many growers arrange to stack the cloches after use between the stretches of cropping plants. This strip method of cropping creates cultivation problems as the cloches get in the way. Trolleys, or platforms on tractor-mounted rear loaders, can also be adapted to carry cloches away from the cropping area. It is normal to find that at least 10 per cent of the glass needs replacing each year.

Cloches, particularly if carelessly stacked, become dirty and will need occasional scrubbing with warm water and a detergent. This is a laborious job, and it is well worthwhile to stack the cloches on polythene covered ground clear of trees and shade to reduce the necessity for cleaning.

Polythene Tunnels*

This method of protection consists of continuous tunnels made of polythene film (150 gauge) suspended on wire hoops and kept down by tying to pegs at each end and to the hoops (Plate V, top). The polythene shields the plants from the wind. It is extremely transparent to solar radiation so that all the heat from the sun penetrates the polythene and warms the plants and ground. Unfortunately, polythene also permits the loss of this heat from the soil and plants during the night. Glass, which does not allow so much re-radiation through the cloche, is therefore much warmer at night. Thus during cold nights, following sunny still days, temperatures under glass cloches will be higher than those under polythene tunnels. But during sunny days, polythene tunnels, if kept closed, will often be much warmer than glass cloches where the spaces between the cloches permit the escape of warm air; sometimes the tunnels become much too hot.

Because of this difference between the properties of glass and polythene, plants growing under cloches are less liable to frost damage than plants under polythene, and this is a point that should be remembered if polythene is used to forward plants in areas where late spring frosts often occur.

Size of Tunnels

The size of wire hoop and width of polythene film to be used will depend largely on planting time and, to a lesser extent, on season, variety and soil for one-year-old plants. Tunnels can also be used to protect autumn cropping varieties and over single, or matted, rows of two-year-old plants.

It is important to assess the expected plant size as accurately as possible since the tunnel should be made just to enclose the plant. If the tunnel is too big there will be unnecessary expense and extra wind susceptibility. But if the tunnels are too small the leaves, flowers and fruit will press against the polythene and may get damaged, particularly if the polythene is opened and lowered for ventilation.

---

* See also Short Term Leaflet 78, *Continuous Polythene Tunnels for Early Strawberry Production*.

On poor soils and for runners planted late September a small tunnel to support 40 in. polythene will suffice. For runners planted at the recommended time of late August to early September in southern England, a medium sized tunnel supporting 48 in. polythene should provide adequate cover. On free-growing soil and to cover plants expected to grow rather strongly, or a double row, or for two-year-old plants, it may be best to use a rather larger hoop and 52 to 54 in. polythene.

Construction of Hoops

Gauge 6 or 8 fully galvanized wire should be used, preferably high tensile wire (50 ton sq. in.) since this makes a more resilient hoop able to withstand pressure and keep a good shape. The wire is best purchased already cut into the required lengths as this avoids the problem of unwinding coils. It is therefore important first to determine the exact size of tunnel and hence the length of wire needed. The length of hoop to be inserted in the ground should not be less than 8 to 9 in. On stone-free soils 10 to 11 in. can be used, particularly for the larger tunnels, and will be less inclined to lift under windy conditions. On stony ground it may not be possible to use long lengths in the soil (Fig. 1).

The ring on each side of the hoop, facing outwards, is made by holding the correct length for the leg end of the wire in a vice and pulling the wire round a small piece of tubing of ½ in. outside diameter. If a large number of hoops are being made, a small jig can be made to aid this bending. Each ring requires 3 to 3½ in. of wire. The wire can then be bent over a suitable barrel, or arrangement of nails in a board, to produce the requisite half circle. The curve should be as smooth as possible to avoid damage to the polythene.

Suggested sizes for tunnel hoops (in.)

	Small	Medium	Large*
Polythene width	40–42	48	50–52
Legs in soil	8	9	10
Height of tunnel	11	12	15
Width of tunnel	18	24	24
Total wire needed	56	64	73
Number of hoops per cwt g. 8 (approx.)	350	300	270

* This length of wire is sufficient to make a hoop 36 in. wide, suitable for a double row of plants, if the hoop is bent to give a curve only 12 in. high, and 52 in. polythene can be used.

If a large number of hoops are to be made, it is worth devising a more elaborate template to hold the wire in the requisite curved loop between two curved pieces on a metal plate. The rings can then be made simultaneously by using a left and right-handed jig to bend the wire around two small iron pegs on the template. Using this type of template the hoops can be made extremely rapidly, whereas using a vice and a single jig, or bare hands, to shape the hooks allows only about 20 to 30 hoops to be made per hour, from pre-cut wire.

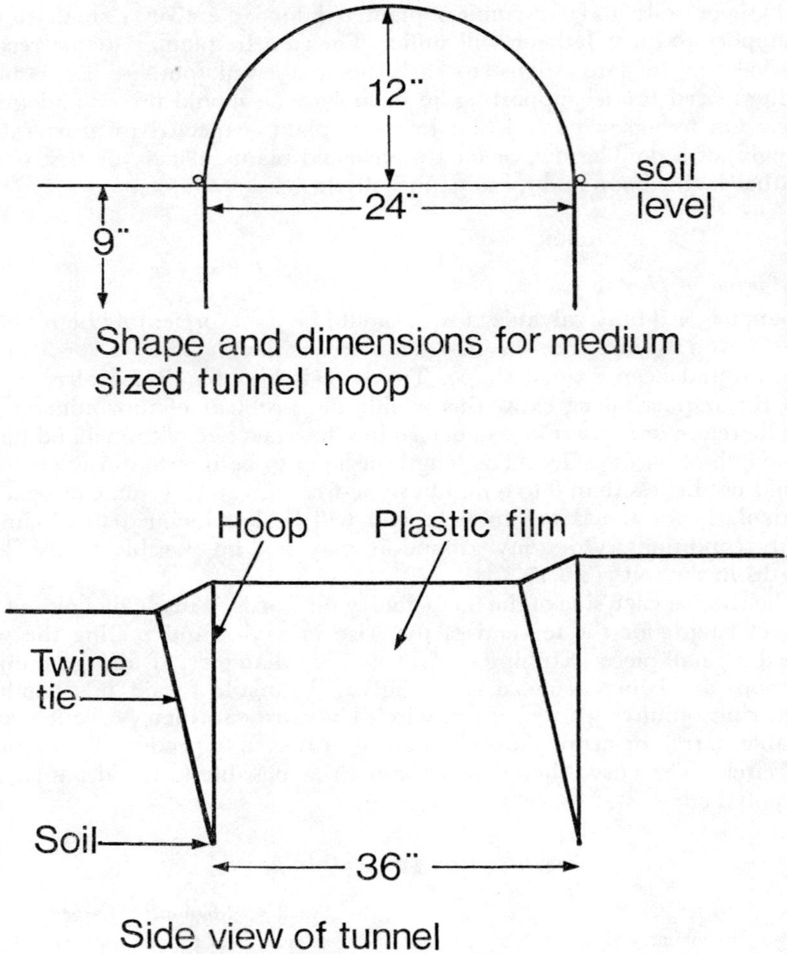

Fig. 1. Construction of hoops

Use of Tunnels

The hoops are placed in position in the field anytime during the winter. Herbicides must be used for systems of protected covering since they remove the need for cultivation. The hoops must be put out in straight lines to allow for a tight compact tunnel. On a small scale, a line can be used down one side of each row at the correct half hoop distance from the centre of the row, and the hoops can be inserted along the line using a stick, or by eye, to gauge the distance apart.

On larger areas and particularly where a tractor and planter have been used to plant two rows, it is best to use the tractor to mark out on each side of the plant rows. A tractor on high clearance row crop wheels is useful as it will clear the hoops after insertion and can be used to run out the polythene; it can also be used for spraying and for fruit cartage. The hoops, tied in bundles of 20 to 30, are placed out of the correct distance over the field ahead of the workers pushing them in the soil.

The hoops must be positioned not further than 4 ft apart and, in exposed or windy areas, many growers use them at 3½, 3 or even 2¾ ft apart to stop wind damage. Two hoops are inserted at the end of each row to take the additional strain at the achorage point; they should be inserted 6 in. apart in the soil and tied together at the top.

The polythene can be placed in position from January onwards. If a large area has to be covered, a mid-January start may be necessary in order to complete the job by early March. The longer the polythene is out the more likely it is to become weather damaged. But sunny days in February and March can give useful soil temperature increases and covering before the end of February is recommended.

The roll of polythene is secured at the end of each row either by tying to a wooden, or iron, stake, or by burying a length in the soil. The polythene is then drawn over the hoops, down the row. On a small scale, the roll of polythene can be carried on a broom handle by two workers. A small hand trolley has been devised to improve on this method. The roll of polythene is held fixed on to a spindle which is free to revolve but can be fixed. The spindle, adjustable for the size of roll used, is supported on a framework on two small wheels sufficiently high to clear the wire hoops. The worker pulls the trolley thus unrolling the polythene. He should not get more than 20 to 30 ft ahead of the person tying down or the polythene may shift with the wind. The spindle is fixed and a good pull taken on the trolley to tension the film over the hoops, which is then eased down over each hoop and secured. For a single-handed application the trolley handles can be fitted with two spikes which can be jammed in the soil to hold the trolley in position while the grower goes back down the row and secures the film to each hoop, starting at the far end.

On a large scale, over 2 or 3 acres, it is worthwhile devising a system of holding two rolls of polythene on a tractor. The tractor tensions the polythene and two rows are covered at one time. Polythene is sold in rolls of 250, 750, 1,000 and 2,000 ft run and the heavier rolls cannot be easily handled without a tractor or trolley.

The polythene is secured to each hoop using polypropylene binder twine. This should be prepared in the correct lengths required with a simple knotted loop at each end. The twine is slipped over one wire ring, pulled over the polythene, drawing it some 1 or 2 in. down below the height of the hoop, and the other end slipped over the other ring. This method is preferable to cutting and tying the twine in the field, a difficult job under cold winter conditions. The correct lengths of twine can most easily be cut by winding the twine over two nails set in a board. The distance apart of the nails must be carefully determined to allow for the two knotted loops at each end. This job can be done when conditions do not permit outside work.

Polypropylene binder twine has replaced g.14 wire which was used previously to secure the tunnels. Twine is more easily cut away with the polythene at the end of the season and dumped. The hoops are collected and stored. Polythene is not usually used a second year; it deteriorates in sunlight and gets dirty.

In windy areas there are several methods of preventing lifting of the hoops and damage to the tunnels. Cross ties of twine or lengths of old polythene secured to wooden pegs every sixth row are helpful and can be readily released for passage of spray machinery or pickers, or extra hoops may be used over the outside of the tunnel. In very windy areas more hoops may have to be used per acre, placing them closer in the row. In the south-west, wires are seamed into the edges of the film and secured to the ground by additional pegs. Straw bales can be placed along the rows to act as a low level windbreak.

Ventilating Tunnels

The polythene can be pushed up between the hoops and ties, allowing ready access to the plants for spraying or picking, a big advantage over glass cloches. Because properly erected tunnels are tight to the ground and continuous, very high maximum temperatures may be reached on sunny days in late April and May, higher than under glass cloches where the warm air escapes through the spaces between the cloches. Without ventilation the plants become drawn and abnormal in growth and more susceptible to mildew infection. Many growers endeavour to ventilate the tunnels during the whole growing, or just the blossoming, period by pushing the polythene up on one or both sides on hot days and pulling down again in the early afternoon. Other growers open the tunnels entirely when about 10 to 20 per cent of the flowers are open and leave the tunnels open until most of the flowers have set. The polythene is then pulled back to the ground or to within 6 in. of the ground. This method, which helps to ensure natural conditions during pollination and fertilization, is only possible in naturally warm areas where frosts are not likely to occur.

It is obvious that the best use of polythene is obtained by opening as necessary to avoid high temperatures and closing again to trap as much heat as possible for the night. But this involves much hand labour, which may be practical for the smallholder doing his own work but not when employing labour. It takes about 8 man-hours to push up on both sides an acre of tunnels. Growers have made short, rubber covered, flat hooks to help them pull up the polythene. Pushing down the polythene takes longer than opening.

Under certain conditions a film of moisture forms on the underside of the polythene film and this is very valuable as it causes the polythene to behave more like glass and to retain more warmth at night. Although the use of tunnels is not recommended in areas liable to spring frosts, it is possible to use water sprinklers designed for frost protection over tunnels on radiation frost nights, and the film of ice so formed should protect the plants from frost damage.

CULTURE UNDER CLOCHES AND TUNNELS

Growing under protection is exactly the same as in the open, and in general the heavy cropping varieties are preferred, even if not the earliest (see section on Varieties). A normal aphicide spray will be needed some two or three weeks before that for field plants. *Botrytis;* rots and mildew may not be such a problem on carefully grown plants under protection. Great care is needed when applying fungicide sprays to protected plants as they are soft in growth and more liable to scorch than unprotected plants.

One of the big advantages of tunnels is that they can be easily opened to allow overhead irrigation. If reasonable rainfall occurs during warm days it may be worthwhile opening the tunnels. Irrigation should finish about 4 p.m. and the cloches or tunnels closed in order to trap sun heat for the night, and to dry out the flowers and foliage.

Overhead irrigation is somewhat difficult through glass cloches, but experience shows that in most years the soil becomes rather dry during picking and that irrigation of cloches and tunnels is very beneficial. If the cloches can be opened, by removal of ventilating panes, overhead irrigation from sprinkler or pipelines will reach the plants sufficiently well. Alternatively, trickle irrigation lines can be laid down the plant rows before cloching and, although expensive, give an excellent system permitting more frequent watering, without wetting the plants. PVC tubing, $\frac{1}{2}$ in. internal diameter with $\frac{1}{32}$ in. holes drilled to come between the plants, is another method that permits of low level watering without removing the cloches. The cost (1969) is about 7d. per ft run.

Because of the high value of protected strawberries, the plants should be strawed if there is any possibility of dirt sticking to the berries. An alternative is to use clear 24 in. polythene placed over the plants on the soil as the cloches or hoops are put out, but it can only be used if an efficient herbicide programme is followed. The polythene has to be slit over each plant and the leaves pulled through. The polythene keeps the fruit clean and gives a small soil temperature increase; it should also help retain moisture. Black polythene is not recommended since it slightly lowers the soil temperature.

Protected strawberries require careful handling and picking. After picking is finished, if the plants are to be retained to crop the following year, it is important to remove the covers, defoliate or clean up the plants, provide irrigation if necessary, and to spray against red spider or strawberry mites, both of which can be troublesome on plants grown under cover.

MATERIALS AND COSTS

The Short Term Leaflet 78 gives details of costs for tunnel construction and these will vary depending on current costs for materials. The quantities required for an acre of tunnels covering rows 3 ft apart, with hoops $3\frac{1}{2}$ ft apart, are approximately:

4,200 hoops 8g wire requiring 14 cwt wire;
5,000 yd of 48 in. polythene;
7,000 yd or 4 rolls of polypropylene baling twine.

It is estimated that the labour to put out the hoops and polythene, using a trolley or tractor, and to remove the tunnels at the end of the season and to give some ventilation is at least 100 hours per acre, additional to the normal labour requirements, but this will be increased if considerable attention is paid to ventilation.

Cloches to cover an acre of plants in 3 ft rows would cost (1969 prices) from £1,800 to £3,000 depending on the type of cloche. A cost of £20 to £30 for replacement glass is usually incurred each year. The handling of the cloches, placing, ventilation, and removal and stacking at the end of the season is more laborious than the use of tunnels and will take at least 150 to 200 hours per acre additional to normal labour requirements.

Late and Perpetual Varieties

THE season for summer fruiting varieties is from mid-May until mid-June for protected plants and mid-June until the end of July for the main crop. The season can be lengthened by perpetual varieties, ripening during August, and then the autumn crop, from especially treated plants of Redgauntlet, follows until late October.

Fruit ripe after the main June-July strawberry season has finished never makes as high a price as the early fruit before the main season. However, there has always been an interest in the so-called perpetual strawberries. These varieties have the capacity to initiate flower trusses during the warmer, longer days of summer and hence flower and fruit from the normal season until the autumn. Because of the long season, there is never a good weight of fruit to be harvested at any one time and picking costs are high.

The capacity for a long flowering season, or for the production of flushes of flowers in cycles throughout the growing season, is seen in many strawberries and, because of the present greater interest in out-of-season production, plant breeders have been paying more attention to the commercial possibilities of these plants. The extreme case can often be found when a seedling produces so many flower trusses that there are few sites left for runners to form. Such plants are obviously of no commercial value because they cannot be easily propagated. Other plants, in addition to producing flowers continually on the mother plant, will also rapidly initiate flowers on the runners so that fruit is also formed on these.

Some perpetual varieties have been grown for several years, the French variety Sans Rivale being the most widely known, but mainly by amateur gardeners. Various recently introduced English and foreign varieties are being tested, but none is outstanding, and now that the perpetuals have to compete with the large sized, autumn fruit of Redgauntlet they are even less likely to be of commercial interest, but might be tried where August fruit is specifically required.

AUTUMN FRUIT FROM REDGAUNTLET

When first introduced, this variety was considered a maincrop, mid-season to late variety. However, it will also flower freely in the late summer, and if plants are protected in the spring, under cloches or tunnels, the flowers produced during July and August will give worthwhile crops in September and October, particularly if protected again. Autumn cropping after spring tunnel protection is less certain than after cloching.

The capacity for second flower initiation in Redgauntlet also occurs on unprotected field plants, but the second fruit generally come so late that it seldom matures, except in exceptionally warm autumns such as 1964.

It has also recently been noted that cold stored runners of Redgauntlet, if grown under tunnels and cloches, are able to produce plenty of flowers somewhat earlier than from fresh dug runners, and will nearly provide a sequence of fruit from May until November.

Other varieties, Cambridge Vigour, Cambridge Rival and Talisman, will also initiate flowers and produce some autumn fruit, but only to a very limited extent and dependent upon weather conditions.

Ripe autumn fruit of Redgauntlet is normally seen only in the south and south-west of England, following warm, early springs and good light conditions. Autumn fruit seldom ripens satisfactorily in the Midlands or East Anglia.

Sufficient autumn fruit from Redgauntlet is now being supplied to the markets during September until early November for a reasonable price to be maintained. However, during early August, when the last of the field strawberries are being harvested from late areas, the price is not so good and it appears that a short break between the two crops is desired by the market. Thus the grower for the wholesale market should aim to obtain autumn, or second crop, strawberries from late August onwards.

If strawberries are being sold retail, or in holiday areas, it is better not to defoliate Redgauntlet after fruiting but to allow the plants to continue flowering and cropping during July, August and until the autumn frosts.

Treatment of Redgauntlet for Autumn Cropping

After the protected crop has been cleared, by mid-June, it is essential not to neglect the plants. The cloches and tunnels should be removed and the leaves may be mown off. In addition to removing old leaves, this will also destroy any flowers that would produce fruit during August. Recent experiments are showing that unmown plants will also crop well and this is helpful as tunnel hoops could then be left in position. Irrigation, essential in most years, must be applied to encourage new growth. A light dressing of 20 to 40 units of nitrogen may be given to help regrowth on poor soils. If herbicides have been used, the soil should be clean and simazine can be applied again, after irrigation (page 39). If the alleyways are weedy, carefully directed sprays of paraquat can be used to burn off the weed growth prior to simazine application. If straw or plastic mulch has been used, this can be left down to protect the second crop of berries. Runners should also be removed.

Insufficient experimental evidence is available to show whether recloching or replacement of tunnels during August is desirable. During hot summers these plants are very liable to infestation by both red spider and strawberry mites, and protection put on too early may encourage these pests and shrivel the plants.

As the plants come into flower, protective sprays against *Botrytis* are necessary. It is suggested that the plants should be recloched and tunnels replaced, depending on weather conditions and labour availability, during September. In very warm autumns as in 1964 it may be possible to harvest good quality crop without protection, but generally night temperatures are too low for good fruit formation and ripening.

Experiments have shown that more autumn fruit is obtained from Redgauntlet plants planted early the previous year in August, or July if cold stored, than those planted later. The crop is less than that in the spring, usually 4 to 6 oz of fruit will be picked in the autumn for each pound in the spring, provided the plants regrow well and are carefully tended.

After the autumn crop the plants can be ploughed in, or they can be retained to crop normally in the following spring as two-year-old plants.

Because two-year-old, and older, plants are slightly later in flowering and seldom protected, it is not common to find good autumn crops produced on these older plants except in exceptionally warm autumns, or if they have been protected in the spring, when they can give good autumn crops.

Autumn fruit is normally of a good size and quality. It should be carefully graded and sold in $\frac{1}{4}$ or $\frac{1}{2}$ lb punnets. The later fruit is much smaller as temperatures fall and picking should stop, whether or not all fruit has been cleared, when quality is so poor that prices are not remunerative, generally in early November.

If the plants are to be retained for the second year, they should not be defoliated again after the autumn crop, since this would check the growth of the flower trusses already initiated for the crop the next year.

Pest and Disease Control

THE great importance of planting only healthy runners has been stressed throughout this Bulletin. In addition, routine sprays for the control of aphids and grey mould are generally essential to maintain healthy plants and to ensure rot-free fruits. Control measures may also be needed for other specific pests or disease, but pesticides should never be applied unnecessarily, particularly those which will kill beneficial insects as well as pests.

The Ministry of Agriculture publishes early each year a useful, free booklet (Agricultural Chemicals Approval Scheme—List of Approved Products)* giving details of proprietary chemicals which have been approved for specific purposes and these pesticides should always be chosen. All approved chemicals carry on their labels a special mark as shown in the diagram on page 74. The manufacturers' recommendations about use, mixing, application and safety measures must be strictly followed. To ensure that the harvested strawberries do not contain any harmful pesticide residues it is important to follow the directions on the strength and frequency of application, and to observe the minimum interval between the last application and anticipated harvest date.

All chemicals should be used with caution and stored in a safe locked place. Empty containers should be carefully washed out, beaten flat if necessary, and disposed of safely. Sprays should not be allowed to contaminate ponds or waterways. Spray operators should wash off any concentrate that may fall on their skin. *There are statutory obligations affecting employers and employees who use certain poisonous substances and these special regulations must be carefully observed.* A leaflet 'The Safe Use of Poisonous Chemicals on the Farm'* is available. Growers should consult their local N.A.A.S. Horticultural Adviser if in doubt about the identification of pests or diseases and their control. The importance of specific pests or diseases varies according to the locality, and useful spray programme guides, prepared by the N.A.A.S. Regional specialists, are available on request. Routine checking and careful observations, together with speedy and adequate control measures should help to prevent serious, or persistent, problems occurring.

---

* See Precautions page 73.

Birds, Rabbits and Hares

Birds can cause serious damage to ripening fruit and may be particularly troublesome on small areas, or when damaging the early ripening berries. Main offenders are blackbirds, thrushes and linnets. Carbide banger guns may give some control but care must be taken to ensure that County byelaws permit their use. Other scaring devices include artificial hawks suspended from helium-filled balloons and various moving, or noise-making, equipment. All are of limited value as birds quickly become accustomed to them. Once the crop is in full harvest with pickers around all day, the loss due to birds is usually less severe. Strawberries under cloches may be protected by closing the ends with wire, or nylon, netting, which keeps out birds but permits ventilation.

Rabbits and particularly hares, will sometimes attack young strawberry plants during the autumn and winter, eating the leaves but generally leaving the crown. They may be particularly troublesome on runner beds. Where severe damage is expected, these areas should be fenced with rabbit-proof netting.

Guide to Strawberry Pests and Diseases
Parts of plant showing symptoms

	Leaves	Flowers	Fruit	Roots	Plants: Wilting	Plants: Stunted
Pest:						
Aphids	+	+				+
Red spider mite	+					
Strawberry mite	+					+
Leaf and stem eelworms	+					+
Tortrix moth	+	+				
Seed beetles			+			
Weevils wingless				+	+	
Blossom weevils } Rhynchites }		+				
Disease:						
Grey mould	+	+	+			
Red core				+		+
Wilt					+	
Mildew	+	+	+			
Zythia fruit rot	+		+			
Yellow edge virus	+					+
Crinkle virus	+					
Arabis mosaic virus	+					
Green petal disease	+	+	+			+
Bronze leaf wilt disease	+				+	

Identification

The accurate identification of pests or diseases is essential if the correct control measures are to be taken, not only to obtain an adequate control but also to avoid the unnecessary use of pesticides. As many materials cannot be used near or during the picking season, preventative or prompt remedial action is often necessary. The above table gives a guide to the important pests and diseases of the strawberry and is followed by a description of the major pests and diseases and their control.

PESTS

Aphids

Several different species of aphid, often called greenfly or plant lice, may occur on strawberry; but only two are of economic importance. These are the strawberry aphid (*Chaetosiphon fragafolii*) and the shallot aphid (*Myzus ascalonicus*).

The strawberry aphid is the main vector (carrier) of yellow edge and crinkle virus diseases (page 68). It causes no obvious feeding injury and is mostly seen on the young unfolded leaves when it is just visible to the naked eye. It is a creamy white aphid characterized by the many knobbed hairs projecting from the head and back. These can easily be seen if the aphid is examined from the side using a magnifying glass.

The shallot aphid is pale greenish-brown, and feeds on the young leaves in the crown. It breeds on the plants throughout the winter and spring, migrating to other host plants in late May. This aphid may cause severe damage after mild winters when high populations produce extensive early leaf distortion and stunting of the plants. The damage resembles, and may be confused with, crinkle virus; the flowers are malformed and little fruit develops.

Control

Control of strawberry aphid is important in order to reduce the spread of virus diseases. The pesticides used will also deal with other aphids which may be present. The systemic organophosphorus pesticides are most effective; these include demeton-S-methyl, dimethoate, demephion, formothion, phosphamidon, mevinphos, ethoate-methyl and granular formulations of disulfoton or phorate. Many of these pesticides are subject to statutory regulations and the necessary precautions must be taken by spray operators.

On outdoor fruiting beds one spray in spring shortly before the blossoms open should give a good control of aphids and reduce the risk of virus spread. On older beds an autumn spray may be worthwhile if aphids have been noticed during the year. One-year plants should be sprayed as soon as picking is over.

If earlier control measures have not been taken and a severe attack of aphids occurs during fruiting, causing plant damage and sticky fruit, mevinphos can be used provided a minimum interval of three days, or malathion provided at least four days, are allowed between spraying and picking. When malathion is used on fruit for processing the interval should be increased to seven days to avoid the risk of taint.

Protected plants under cloches or tunnels should be sprayed in late March. Unsprayed plants frequently become badly infested as the aphids breed rapidly under these sheltered conditions.

Runner beds must be sprayed more frequently to ensure adequate protection from migrant, winged aphids which might be infected with virus; at least three sprays are necessary in mid to late May, June and again early in July.

Red Spider Mite
Advisory Leaflet 226

The species found on strawberries is the glasshouse red spider mite (*Tetranychus urticae*). It overwinters in the adult female stage sheltering in the soil and on the plants, straw and weeds. These mites become active in April or earlier on cloched crops. The tiny red or greenish mites, cast skins and dull white eggs can easily be seen with a magnifying glass on the underside of the leaves where they feed, causing extensive patchy mottled areas. Some varieties, such as Cambridge Favourite, are very susceptible and in some seasons, in the absence of control measures, cloched plants in particular may become heavily infested before picking. The growth becomes poor, the whole plant appears bronzed and sometimes becomes covered with webbing. In some seasons severe damage may also occur on outdoor plants.

Control

The pesticides recommended for aphid control in April should also control red spider mite and prevent a severe build-up before harvest. In some seasons a second spray may be required after harvest. Under cloches, where conditions favour rapid breeding, pre-blossom sprays should be applied as a routine practice.

In 1964 some growers reported difficulty in controlling red spider mite on strawberries with systemic organophosphorus pesticides. It has been confirmed that, on a number of holdings, particularly in the Wisbech, Hampshire and Cheddar strawberry growing areas, strains of mites have developed which are resistant to these pesticides. If poor control is obtained other materials must be used, i.e., tetradifon or dicofol before flowering and, if necessary, after picking. These pesticides will not control aphids and the routine aphicide spray should be applied before flowering.

Strawberry or Tarsonemid Mite
(*Steneotarsonemus fragariae*)

This mite is occasionally troublesome in some areas, particularly in warm summers, and has become more prevalent in recent years. The microscopic mites feed in the young folded leaves, causing roughened rust-coloured areas and a characteristic leaf-edge curl. Later in the season the crown leaves may turn brown and die and the whole plant may lose its fresh green colour, the older leaves being puckered and wrinkled. With severe infestations the plant may become stunted. Damage is unlikely to be obvious until after fruiting. Cloched plants and older beds are particularly liable to attack.

Control

Strawberry mite can be spread on infested runners; runners should be taken only from stocks known to be free of mite. New plantations should also be situated as far as possible from infested beds. Strawberry mite is difficult to control, and the organophosphorus pesticides recommended for red spider mite control will not kill it and may aggravate the problem by killing predators. Endosulfan and endrin will give a good control, and dicofol is also effective although control is usually slower. Sprays should be thoroughly applied after picking using high volume applications. Endosulfan and endrin must not be used on plants which will be cropped again in the autumn, and these should be sprayed with dicofol, provided a minimum interval of seven days is allowed between spraying and picking.

Eelworms or Nematodes

Advisory Leaflet 414

Eelworm damage to the leaves and crown of the plants may be confused with shallot aphid damage but can be distinguished by the absence of aphids and cast skins. Eelworms are very small and cannot usually be seen with the naked eye, or easily with a magnifying glass.

Leaf Eelworms

These species (*Aphelenchoides fragariae* and *A. ritzemabosi*) feed on the young developing leaves, causing damage most noticeable in the spring when growth is slow. Affected leaves may be puckered, distorted and fail to open normally; some leaflets may show greyish or silvery feeding areas on the upper surface towards the base of the midrib. The main crown often becomes blind and many secondary crowns then develop, forming an open-centred plant which crops poorly. Runners from infested plants will also be infested but may look healthy.

Stem Eelworm (Ditylenchus dipsaci)

This eelworm generally causes a marked corrugation of the leaves and a shortening, thickening and distortion of the leaf stalks (Plate VIII, top). There are several races of stem eelworm, differing mainly in the host plants on which they live. It is known that the oat, red clover and narcissus races can infest strawberries. Parsnips, rhubarb, broad beans, vetches and mangolds are also important hosts of this eelworm, as are many weeds.

Strawberries should on no account be grown on land where any of these crops have previously been infested by stem eelworm. Runners should never be taken from infested plants.

Control

To avoid trouble from eelworms it is essential that only clean runners should be planted, as little can be done to clean up infested plants. If only small numbers of plants are affected in fruting beds, these can be dug out and burnt to reduce the risk of spread. There are no satisfactory control measures for heavily infested beds.

In areas where bulb crops are frequently grown in rotation with early strawberries, soil fumigation with D-D, although expensive, may be essential. Attacked crops should be lifted as soon as possible and the ground left free of strawberries overwinter. The soil should be treated with D-D the following summer (page 61).

Free-living Soil Eelworms

The dagger eelworm *Xiphinema diversicaudatum* a soil-inhabiting, root-feeding species, is a locally important pest of strawberries because it transmits arabis mosaic virus (page 69). The eelworms do not themselves seriously damage the plants unless unusually large numbers are present, but if they are carrying the virus they will infect the plants causing severe crop loss. *X. diversicaudatum* can also transmit strawberry latent ringspot virus and in Scotland, certain species of needle eelworm (*Longidorus spp.*) are vectors of strains of raspberry ringspot and tomato black ring viruses which infect strawberry. Both kinds of eelworm vector can be controlled with D-D soil fumigant.

Control

The dagger eelworm is very sensitive to D-D and is controlled by soil injection carried out at least six weeks but not more than six months before planting. Autumn application has been shown to give good control of the eelworm

D-D Soil Fumigation

D-D is used at a rate of 33 gal per acre. It is sold by weight, 1 gal weighing 12 lb. This fumigant will kill eelworms and other soil pests, weed seeds and plants; thus it cannot be used after the crop is planted.

The chemical must be injected at a depth of about 8 in. into well-worked, moist soil where it vaporizes; the fumes penetrate the soil, killing the eelworms. D-D injection can never completely eliminate the pest as eelworms in the deeper layers can reinfest the treated soil.

After injection the surface should be rolled or sealed by thorough watering. The treated area must be completely free of D-D vapour before strawberries are planted. Three weeks after injection the soil must be thoroughly cultivated followed by a second cultivation two weeks later. Planting *six weeks* after injection should be safe provided that the soil has been well cultivated.

D-D is a very corrosive liquid and care must be taken when handling it and in cleaning machinery after use. A hand injector can be used for small areas; the single spike is pressed into the ground at 15-in. centres in staggered rows, the plunger then delivers a pre-set amount of liquid at each point.

A tractor-mounted machine can be used for large areas. The D-D is gravity fed from a tank through polythene tubes to hollow injector tines. The biggest problems are to control the flow of D-D and to obtain deep enough penetration of the tines.

Tortrix Moths

A number of species occur but the strawberry tortrix (*Acleris comariana*) is the most commonly found. The caterpillars feed on the leaves and flowers, webbing the leaves together or feeding in a silken web on the underside of the leaf. They emerge in April or early May from eggs which have over-wintered on the plants. The young caterpillars are white with black heads later becoming green with a shiny yellow head. The adult moth emerges during late June and July and eggs are laid on leaf stalks and stipules. A second brood may be found on the plants in August and September. This pest is considered to be important in the strawberry growing areas near Wisbech and in west Norfolk.

Control

DDT, which was commonly used to control this pest, may encourage the build-up of red spider mite so that it should not be used as a routine, but only where damage is expected and the first caterpillars seen. Where the pest is a problem, azinphos-methyl or DDT should be applied in April when damage is first seen. DDT will not give a satisfactory control after the leaves have been webbed together, thus protecting the caterpillars. A second spray may sometimes be required in late summer.

Strawberry Seed Beetles (*Harpalus rufipes*)

Adult beetles damage both ripe and unripe fruit by removing the seeds, causing superficial injury to the flesh. However, damage to strawberries by linnets is very often confused with that caused by beetles. It is important to distinguish between the two types of injury so as to avoid unnecessary use of pesticides. The main difference is that linnets remove the seed cleanly, usually only from the upper surface of exposed fruit. Seed beetles injure the flesh around the seed and may attack any part of the fruit, particularly the underside next to the soil.

The adult beetles are a dull black colour with yellowish brown legs. They overwinter in sheltered places and move into the beds soon after strawing. Grass headlands, box stacks or nearby hedges and woodland provide ideal shelter for overwintering beetles and should be avoided whenever possible when choosing a site. A weed-free area around the fruiting beds will reduce the winter shelter. Other beetles (*Feronia spp.*), injure the fruit by eating large holes in the flesh; this is sometimes confused with slug damage.

Control

Recent work has shown that methiocarb slug pellets can also give some control of ground beetle damage. Where attacks are expected an application before the fruit ripens may be worthwhile although complete protection is unlikely with these baits.

If ground beetle attacks occur during fruiting, a poison bait made with 1 pt 60 per cent malathion concentrate, 8 pt water and 40 lb of crushed oats per acre will give a useful kill of beetles and will reduce further damage. The bait should be placed between the rows in small heaps at about 1-yd intervals; it remains effective for about a week but further applications can be made if required.

Weevils

Advisory Leaflet 57

Wingless Weevils

Several species may occur differing with locality. Vine weevil (*Otiorhynchus sulcatus*) and strawberry root weevil (*O. rugostriatus*) are two commonly found species. The creamy white, legless grubs feed on the root tips in the winter and spring, finally excavating deep tunnels in the base of the crown and so causing complete collapse of the plants. Attacked plants are usually isolated and collapse quite suddenly, so this damage need not be confused with verticillium wilt. The adults feed on the foliage at night during May and June but damage is unlikely to be important.

Another weevil, the red legged weevil (*O. clavipes*) is a locally serious pest in parts of the south-west, causing root damage. There are two main periods of adult activity, the first in spring and again in mid-June and late August.

Control

No satisfactory control measures can be taken against these well-protected weevil grubs when the attack is first seen during flowering or fruiting. To prevent the infestation spreading, apply DDT dust (100 lb 5 per cent dust per acre) to the crown of the plants and surrounding collar of soil in mid-August, for vine and strawberry weevil, or in early September for the red-legged weevil. This will kill newly hatched grubs before they penetrate deeply into the soil. Alternatively a DDT drench (3 pt 15 per cent DDT per 100 gal at 1 pt per plant) can be used.

Where land is known to be infested with these pests, it should be treated with *gamma*-BHC worked into the top 4 in. of soil two or three weeks before planting. *Gamma*-BHC should not be used on land to be cropped within 18 months with potatoes, onions, carrots or other root vegetables for human consumption because of the risk of taint.

Strawberry Rhynchites (*Caenorhinus germanicus*) and Strawberry Blossom Weevil (*Anthonomus rubi*)

Both species destroy strawberry blossoms and may occur on the plants at the same time. The adult weevils are small and seldom seen. The damage caused by each weevil differs. Strawberry blossom weevil is generally more common than rhynchites; it is sometimes called the 'elephant beetle' or 'needle bug' and is greyish black. It lays its eggs in the unopened flower buds and partially severs the individual flower stalk causing the flower to wither. The strawberry rhynchites is a small blue-green weevil which punctures, and lays eggs in, the main stem of the truss and young leaves, causing the whole truss to wilt and die.

Control

Damage by strawberry blossom weevil can be reduced by spraying with DDT or TDE or preferably dusting with DDT at the early green flower-bud stage or when damage is first seen. A second application may be necessary 10 days later. To control rhynchites, DDT sprays or dust should be applied in mid- to late April.

Leafhoppers

Leafhoppers (*Aphrodes spp.* and *Euscelis spp.*) are the vectors of a disease that causes the green petal condition in strawberry (page 70). The leafhoppers themselves do no damage to the plants. They feed on a wide range of plants and are active for a considerable period, making chemical control measures difficult and of doubtful value. Wide, clean headlands around the runner beds may help to prevent the spread of hoppers into the field.

FUNGAL DISEASES

Grey Mould

Also see Short Term Leaflet 64

Fruit rot or grey mould, caused by the fungus *Botrytis cinerea,* is a widespread and damaging disease of cultivated strawberry. The disease causes rotting of the green and ripening fruits, quickly reducing them to a soft mass covered with grey dusty spores of the fungus and, under humid weather conditions, can devastate the crop. The disease is also responsible for a reduction in yield due to *Botrytis*-induced abortion of flowers. Although there is considerable varietal variation in susceptibility (see section on Varieties) this disease is mainly influenced by seasonal weather conditions.

Sources of *Botrytis* are very common and virtually impossible to eradicate. The fungus is an inhabitant of dead plant tissue from which it can attack living plants. Old dead strawberry leaves and leaf stalks provide excellent material for the fungus to live on and to produce spores, if given favourable weather conditions. The greatest danger comes from early infection of flower parts, such as the petals, sepals (calyx), stamens and receptacle by airborne spores, or by contact with infected material. Infection at this stage may remain latent and then cause fruit rotting at harvest time. Fruit rotting also occurs by direct infection of ripening fruits by spores, or by contact of fruit with infected flower parts, or other rotting berries.

Control

Good control of grey mould can only be achieved by ensuring a thorough cover of fungicide, particularly during the blossoming period. Because of the long flowering period of the strawberry, three or four sprays are necessary, applied at the following times:

1. at early flowering; when first flowers are open on most plants (Plate VII, top);
2. at full blossom; when most of the flowers are open on most of the plants (Plate VII, middle);
3. at petal fall; when most petals have fallen on most of the plants (Plate VII, bottom);
4. again, 10–14 days later, where possible.

Later spray applications will protect the late opening flowers but do raise practical problems of spray residues on already developed fruits, especially when these are intended for canning or quick freezing, as well as the risk of damage by the passage of spray machinery.

The most efficient fungicide is dichlofluanid. Experiments have shown this to be more effective than thiram, and thiram more effective than captan. When using any of these fungicides one should ensure that the stipulated interval between the last spray and harvest is strictly adhered to. The efficiency of fungicides depends on their correct application: this is particularly so in the control of *Botrytis* and very thorough wetting of all parts of the plant, especially the flower trusses, is essential if good control is to be achieved. Low volume spraying is less satisfactory than high volume spraying and on established plants 100–200 gal/acre are necessary.

In tests to date, dichlofluanid has not caused taint in the processed (canned) crop. Thiram and captan are liable to make fruit unsuitable for processing. Captan must not be used on strawberries for canning and thiram not for canning, quick freezing or jam. Dichlofluanid can be mixed with dinocap for mildew control but sometimes causes damage.

Dichofluanid is not recommended for use on protected strawberries because of the danger of leaf scorch. Thiram under dry conditions can also cause leaf scorch to protected strawberries but the risk is lessened if liquid thiram preparations are used. Captan is safe, but less effective in controlling *Botrytis*. Small, one-year-old protected strawberries are less likely to attack and spraying is less important. It is probable that thick, dense plants and particularly those in overcrowded, matted rows are most liable to *Botrytis* infection. Crowded plantations require careful spraying and the use of more spray material than smaller, younger, or more widely spaced plants.

Attention to plantation hygiene can do much to lower the level of the disease. Where infection has already begun, rotted fruits should be removed from the bed during picking, rather than left on the plant, or thrown down in the rows. Burning-over at the end of the fruiting season is a convenient way of cleaning up a strawberry bed; this destroys the straw and the leaf debris, and destroys large numbers of *Botrytis* resting bodies (sclerotia).

Whilst all varieties are susceptible to grey mould, Cambridge Rival, Cambridge Prizewinner, Redgauntlet and Gorella are the least susceptible.

Red Core

Advisory Leaflet 410

Red core disease has been recorded in all the main strawberry-growing districts. The fungus, *Phytophthora fragariae*, lives in the soil and attacks the roots of strawberry plants causing them to rot. Infected plants are more often seen where drainage is poor and the incidence of red core is always worse following a wet autumn and winter. The effects of the disease are normally first noticed in late spring, or early summer, when diseased plants look as if they are suffering from drought; the plants are dwarfed and may wilt and die, particularly in dry weather. Their leaves are often dull bluish-green and the older leaves are sometimes tinged with brown or red. Occasional plants may wilt suddenly without previously showing any of these symptoms.

If the roots of an infected plant are examined in spring it will be found that the white feeding rootlets have rotted away, so that only the white fleshy main roots together with the old brown, rather woody, roots remain. The white fleshy roots may be brown and rotten for varying distances from their tips. These roots have been described as 'rat's tail' roots. If such roots of an infected plant are cut lengthwise the characteristic red discoloration of the central tissue (stele) may be seen (Plate VIII, bottom right). This is the red core from which the disease takes its name. However, it is possible to confirm red core infection by microscopic examination, even when characteristic red core symptoms in the roots are not visible.

The disease is spread from one area to another and from field to field on infected runners. Contamination of land can also result from the carrying of infected soil on implements from field to field. Once the disease has been introduced into a field its development depends very largely on soil moisture conditions during autumn and spring. Under conditions of bad drainage, or high winter rainfall, the fungus readily invades strawberry roots to give the characteristic symptoms of the disease. In the infected tissues of the root the fungus forms thick-walled resting spores. When the infected roots rot away, these resting spores remain in the soil and apparently survive indefinitely.

Control

It is very important to prevent the disease being introduced into clean land and every effort should be made to ensure that runners are free from infection by planting only certified runners.

There is at present no cure for this disease and rotation of other crops, however lengthy, will not ensure that the disease has been eliminated. Much can be done to reduce losses from red core by attention to, and improvement of field drainage. In some infected areas plants are grown on ridges and this appears to aid growth, presumably by improving surface drainage. Early planting, during August and early September, encourages root growth and provides a larger root system, better able to tolerate infection before onset of winter than does later planting. Replanting each year with strong healthy runners is advised on fields where the disease is endemic. Plants grown under cloches and tunnels may crop satisfactorily although infected, as the protection keeps the roots drier in the spring.

Control of the disease by the use of field resistant varieties is complicated by the existence of many strains or races of the fungus. The races differ in their ability to attack and infect different varieties. A variety which is field resistant in one area may become infected in another area when subjected to a particular strain of the fungus to which it is susceptible. Varieties in the field differ in their resistance to the disease: Templar, Crusader, Talisman, Cambridge Vigour, Cambridge Rival, Redgauntlet, and Gorella are tolerant to some but not all strains of the fungus; Merton Princess and Cambridge Prizewinner are susceptible. No variety is field resistant to all strains of the red core fungus, nor is there any true immunity.

Under the Red Core Disease of Strawberry Plants Order 1957 anyone who knows, or suspects, that plants growing on his land are infected must notify the Ministry of Agriculture, Fisheries and Food. The sale of infected plants or plants grown on contaminated land is prohibited. Such plants may not be planted except on land declared to be infected.

Verticillium Wilt

This fungus disease is caused by *Verticillium albo-atrum* and *V. dahliae* which live in the soil and under favourable conditions infect the plant causing it to wilt. The disease usually causes greatest loss in maiden plants. If such plants are able to tolerate infection in the maiden year they usually survive to produce a crop in subsequent fruiting years. Yields from infected plants are appreciably below those obtained from healthy plants. The disease can be introduced into clean soil by planting runners infected with verticillium wilt. Wilt is a less damaging disease than red core and its incidence fluctuates, depending on seasonal weather conditions.

This disease is difficult to diagnose correctly in the field and is generally seen in the summer. Infected plants wilt, especially during dry weather, the older leaves turn brown and shrivel and eventually the wilted plant may die. If the crown of an infected plant is cut transversely a brown discoloration may be visible in the wood tissue. Infection is often severe when strawberries follow other susceptible cultivated crops, especially potatoes.

Varieties differ in their susceptibility to wilt, but no present day variety is fully resistant.

Control

Recent experimental work has shown that pre-planting soil injection with chloropicrin will almost completely control wilt for a period of three years. This practice is expensive and is best done by a contractor using specialist equipment. Treatment may be justified for runner beds or for specialist production where rotation is difficult and the disease severe. Under the Agriculture (Poisonous Substances) Regulations 1966, chloropicrin has been scheduled as a Part I substance, this means that full protective clothing and an approved respirator must be used when handling the concentrate.

Mildew

Mildew, caused by the fungus *Sphaerotheca humuli*, is a common disease of cultivated strawberries, particularly during dry periods. Leaves of attacked plants show characteristic dark red coloured blotches usually on the upper surface; later, the margin of such leaves curl inwards to expose the undersides which are covered with a white powdery mildew. Infected plants, showing these characteristic symptoms, appear to be suffering from drought. The fungus also covers the flower-stalks, flowers and fruit with a white powdery mildew. Mildew is always worse under hot dry conditions.

Control

The disease can be effectively controlled by three sprays of dinocap applied at early flowering, at full blossom and petal fall. Dinocap can be added to the sprays for *Botrytis* control. Dinocap sprays sometimes damage petals and mixed dinocap-dichlofluanid sprays may cause some scorching of foliage.

Runner beds should be sprayed at fortnightly intervals from the time new crown leaves develop, but Cambridge Favourite, Redgauntlet and Talisman may not need spraying. Varieties differ in their resistance: Cambridge Favourite, Talisman, Redgauntlet, Merton Princess, and Cambridge Prizewinner are tolerant, whereas Cambridge Vigour, Gorella, Templar, and Crusader are susceptible, Cambridge Vigour particularly so.

Fruit Rot and Leaf Blotch caused by *Zythia fragariae*

Rotting of fruit can sometimes be attributed to attack by the fungus *Zythia fragariae*. The fungus can also cause a leaf spot and black sunken areas on the leaf stalks and stolons. The leaf spotting phase has been troublesome in some seasons, particularly in the Cheddar area. The disease has been most damaging in wet summers when noticeable loss of fruit may occur in the susceptible varieties Redgauntlet and Cambridge Favourite.

Symptoms on the fruit are first seen as shrivelling of the calyx, which later curls upward. The disease progresses into the berry where it causes a slightly sunken, fairly firm brown rot of the shoulder of the fruit. Tiny, flesh-coloured spore bodies may be seen in the rotted area under the calyx and, if the fruit surface is wet, a milky mass of spores may be produced. Attacked fruit soon become infected by *Botrytis* which very quickly masks infection by *Zythia*.

Control

The disease does not appear to be adequately controlled by the fungicides at present used to control *Botrytis*.

VIRUS DISEASES

Aphid-Transmitted Diseases

Advisory Leaflet 530

The common viruses which cause the diseases known as yellow edge and crinkle are spread from plant to plant by the strawberry aphid. Control of the aphid is therefore essential for control of these diseases. Viruses are present throughout the plant as a whole and all parts of the plant and runners are infected.

Yellow Edge Virus

Typical symptoms appear in sensitive varieties during late summer and autumn and consist of general stunting of the plant and yellowing of the leaf margins of the younger leaves. The yellow edge is not sharply defined but merges gradually into the green part of the leaf. The diminished growth of the young leaflets results in a dwarfing of the leaf as a whole, the margins of the leaves tend to turn upwards, the leafstalk is normally short and, in severe cases, the whole plant presents a dwarfed, flattened appearance. During summer, the yellowing of the leaf margins is not apparent. Yellow edge should not be confused with green petal virus (page 70).

Crinkle Virus

Plants infected with mild crinkle virus show faint chlorotic spots which are less numerous than with severe crinkle, and leaf distortion is absent or slight. The symptoms are most pronounced in early summer and in the autumn.

Symptoms of severe crinkle virus diseases are first seen in early summer on the young leaves of an infected plant. These show numerous small yellow (chlorotic) spots which may become red or brown (necrotic). The foliage becomes severely crinkled and distorted, often with red coloration. In severe crinkle the leaf stalks are not noticeably shortened and the plant retains its normal habit in contrast to the flattened, yellow edge infected plant.

Control

Many present day commercial varieties are tolerant of these virus diseases particularly mild crinkle: Cambridge Rival and Cambridge Prizewinner are the most likely to show symptoms. Most commercial varieties are fairly sensitive to yellow edge disease, Redgauntlet showing very marked sensitivity. The implementation of the certification scheme and efficient aphid control have also helped in controlling the spread of these diseases. Runner beds must be frequently treated and inspected and all doubtful plants and runners removed and destroyed. Fruiting beds should be given routine aphicide treatments (page 58) to kill the aphids which spread these diseases. Obviously diseased plants should be removed and burnt. Runners should never be taken from infected plants. New plantings should be sited as far away as possible from older strawberry crops.

Eelworm-Transmitted Diseases

Advisory Leaflets 530 and 564

Four virus diseases of this group affect strawberries: arabis mosaic, strawberry latent ringspot, raspberry ringspot, and tomato black ring. They are transmitted to healthy strawberries by root-feeding eelworms and are prevalent in certain areas.

Arabis mosaic virus is the only one of economic importance in England and Wales. Strawberry varieties differ in their reaction to infection with this virus but in general the leaves are distorted, mottled, or blotched with yellow; some varieties, particularly Cambridge Favourite and Cambridge Prizewinner, are severely stunted. On other varieties the symptoms of arabis mosaic infection can be confused with those caused by crinkle virus, whilst some are only slightly affected; none is immune. No obvious symptoms develop on Redgauntlet. Because of the possible confusion with crinkle and the expense involved in chemical control of arabis mosaic virus it is most important to obtain confirmation of the presence or absence of this disease.

Infection occurs through the root-feeding activities of the eelworm vectors which transfer the virus from other infected strawberries, or from infected weeds. Control of the eelworms is thus essential for the control of this disease. Replanting healthy runners on land known to be infested with infected eelworms must result in further trouble. Infected weeds often show no symptoms and in areas where the disease is known to occur weed control is important.

Control

D-D soil injection (page 61) has given good control of the eelworms and hence the disease. After land has been treated great care should be taken to plant only healthy runners from certified sources. If isolated infected plants showing symptoms of arabis mosaic virus are noted, the plant and surrounding soil should be removed and burnt.

Leafhopper-Transmitted Diseases

Advisory Leaflet 530

These 'viruses'* can affect strawberries, clover, a few ornamentals and some weeds. Leafhoppers carry the diseases to strawberries, but spread is slow and generally only scattered plants are affected.

Green Petal Disease

This 'virus' complex causes very different symptoms depending on the strain of 'virus' and the time of year when infection takes place, and the disease is more prevalent in years following a long, dry summer. In the spring young leaves produced on an infected plant are chlorotic, much reduced in size, and sometimes asymmetrical, and the plant is very distorted and dwarfed. After flowering the older leaves become red or bronze. The whole plant may collapse in summer; should it survive, the young leaves produced in the autumn are small and the plant gradually dies. If the plant flowers the petals are small and sometimes green and the fruit is small, leathery and 'pippy'. Symptoms of the infection develop slowly in an infected plant and sometimes an older plant may have several healthy and several infected crowns. The bronzing and wilt in summer distinguishes this disease from yellow edge, whilst the presence of small flowers, shrivelled fruit and very small chlorotic leaves distinguish it from verticillium wilt. All varieties are susceptible to green petal 'virus' but often do not show the green petal symptom.

Bronze Leaf Wilt Disease

This is caused by the same organism as green petal but gives different symptoms and is most frequently seen on the variety Cambridge Favourite. Infected plants wilt and die very rapidly, sometimes in full fruit; the outer older leaves turn a marked bronze colour. Green petal symptoms are not seen.

Control

Control of the leafhoppers which carry this disease is difficult, since activity of the leafhoppers is dependent on seasonal weather conditions. Hot, dry weather during July/August leads to increased movement of the leafhoppers and disease spread takes place. Symptoms of green petal disease are mostly seen in the year following infection. Since spread of the disease within the plantation is slow and only isolated plants are infected, these diseases are not so important as the aphid-transmitted virus diseases. Plants showing symptoms in runner beds should be removed complete with their runners. Removal of plants in young fruiting plantations is also worthwhile.

---

* These diseases are now known not to be caused by viruses, but by simple organisms called microplasma.

SPRAY MACHINERY

Tractor Mounted Sprayers*

The choice of spray machinery for strawberries will depend very much on the crops grown on the rest of the holding, since strawberries are seldom the only crop and are not often grown on a big enough acreage to justify expensive specialist tackle. Where a four-wheeled tractor is available, a tractor-mounted machine consisting of spray tank and pump, booms and nozzles may be used. Provided the nozzles do not exceed the output of the pump, any pump will suffice, as very high pressures are not required for spraying strawberries. The normal ground crop spray boom with fixed, down-positioned nozzles is satisfactory for application of herbicides and for aphid sprays in the spring before the plants have grown too dense. However, it is unsuitable for applying fungicides to dense plants since the position of the nozzles and their low volume outputs (usually 25–50 gal/acre) cannot ensure an adequate cover.

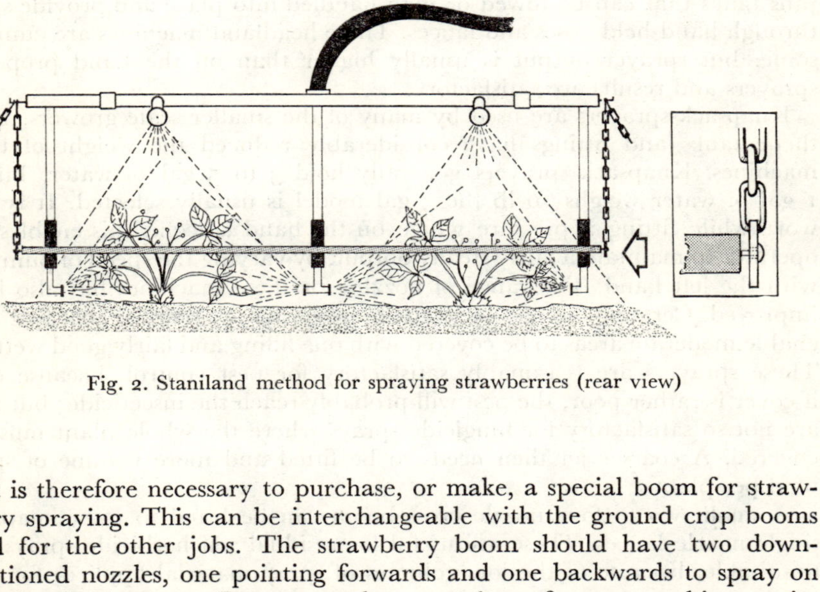

Fig. 2. Staniland method for spraying strawberries (rear view)

It is therefore necessary to purchase, or make, a special boom for strawberry spraying. This can be interchangeable with the ground crop booms used for the other jobs. The strawberry boom should have two down-positioned nozzles, one pointing forwards and one backwards to spray on top of each plant row. Between each row a piece of pressure tubing carries two adjustable nozzles to spray the plant rows sideways from just above soil level. An adjustable, rubber-cushioned rod is positioned to press against the leaves and turn them over as the machine travels down the row, and so help the spray reach the undersurface of the leaves and penetrate the crown.

---

* More detailed information is given in Bulletin 5, *Fruit Spraying Machines* (1967), price 6*s*. *od*., (30p) (post extra), obtainable from H.M.S.O. or through any bookseller.

High volume spraying is recommended at the rate of 100–200 gal/acre, and H.V. type nozzles should be used. The width of boom and hence number of nozzles will depend on the pump output. The tractor P.T.O. and land speeds should be adjusted to give the required volume per acre (see section on Weed Control with Herbicides and Cultivations). The number of rows that can be sprayed at one time will depend on the sprayer output. Where different row spacings are used on one holding the spray nozzle positions will need to be adjustable.

As so many accidents occur through contamination of incompletely cleaned sprayers, it is recommended that a separate machine should be used for hormone, or desiccant type, herbicides. All power sprayers should be well maintained and very thoroughly washed out after use.

Small and Hand Sprayers

On small areas of strawberries where there is no tractor, smaller machines that are quite satisfactory are available. These include hand propelled machines with 5 or 10 gal tanks and a range of small headland sprayers plus tanks that can be towed or man-handled into place and provide spray through hand-held hoses and lances. These headland machines are cumbersome, but sprayer output is usually higher than on the hand propelled sprayers and results are satisfactory.

Knapsack sprayers are used by many of the smaller scale growers. Polythene tanks and fittings have considerably reduced the weight of these machines. Knapsack sprayers generally hold 3 to 5 gal of water, but as 1 gal of water weighs 10 lb the 3 gal model is usually selected. It is well worthwhile fitting a pressure gauge on the hand lance as this enables the operator to maintain a fairly steady output, by varying the speed of pumping with the left hand. The range of nozzles for these machines has also been improved. Ceramic fan jets for low volume sprays are popular, because they enable moderate areas to be covered with one filling and fairly good wetting. These sprayers are reasonably satisfactory for pest control, because even if cover is rather poor, the pest will probably reach the insecticide; but they are not so satisfactory for fungicide sprays where the whole plant must be covered. A coarser jet then needs to be fitted and more volume of spray used per plant.

A small two to four nozzle cross boom can be fitted to most knapsack single nozzle lances. These are useful for residual soil herbicide sprays but unsuitable, because of the low volumes of spray material applied, for pest or disease control.

Spraying should always be carried out under good conditions with the minimum of wind. There is more likely to be damage from spray materials, or the various wetters with which they are formulated, under slow drying conditions. Spraying under dull wet conditions should be avoided. Conversely, very hot dry conditions may cause plants to be more sensitive to spray injury.

The manufacturers' instructions should always be carefully followed. Materials should be accurately measured out and properly mixed and protective clothing worn as specified. Sprayers must be well washed out after use, nozzles and stop ends taken off and clean water pumped through and drained away in a safe place.

Dusters

Some insecticides and fungicides are sold in dust form and both tractor mounted and hand dusters are available. Dusting is, however, not so efficient as spraying, under normal conditions in this country.

PRECAUTIONS

Whenever pesticides are used, read and follow carefully the instructions on the label. Users should also consult the 'Recommendations for Safe Use of Chemical Compounds Used in Agriculture and Food Storage', published by and obtainable free from the Ministry of Agriculture, Fisheries and Food, Food Standards Science and Safety Division, Great Westminster House, Horsefery Road, London S.W.1. Use the pesticides mentioned in this Bulletin with care, especially—

aldrin
azinphos-methyl*
gamma-BHC
chloropicrin
'D-D'
DDT
demeton-S-methyl*
dimethoate
dinocap
diquat
disulfoton*
endosulfan*
endrin*
ethoate-methyl
formothion
lenacil
malathion
mevinphos*
paraquat
phorate*
phosphamidon*
thiram

and wash off any concentrate that falls on the skin. Store new and part-used containers in a safe place under lock and key. Wash out empty containers thoroughly and dispose of them safely. Do not contaminate ponds, ditches or waterways with concentrate, spray or washings.

Protection of Consumers

To ensure that a harvested crop does not contain any harmful pesticide residues, follow the instructions given on the label on the strength and frequency and time of application of a pesticide and observe the minimum intervals between the last application and harvesting.

---

* There are statutory obligations affecting employers and workers who use certain poisonous substances, including azinphos-methyl, demeton-S-methyl, disulfoton, endosulfan, endrin, mevinphos, phorate and phosphamidon. Users of these chemicals are strongly advised to read the Ministry's Leaflet APS/1, 'The Safe Use of Poisonous Chemicals on the Farm', obtainable free from the Ministry of Agriculture, Fisheries and Food (Publications), Government Buildings, Block C, Tolcarne Drive, Pinner, Middlesex, HA5 2DT or any Regional or Divisional Office.

Proprietary products based on chemicals used for pest, disease and weed control can be offically appoved under the Agricultural Chemicals Approved Scheme. It is strongly recommended that approved products should be used. Approval is indicated on the containers by the mark shown here. A list of approved Products is published in February of each year and is obtainable free of charge from the Ministry of Agriculture, Fisheries and Food (Publications). Government Buildings, Block C, Tolcarne Drive, Pinner, Middlesex, HA5 2DT or any Regional or Divisional Office.

AGRICULTURAL CHEMICALS APPROVAL SCHEME

Harvesting

HARVESTING is the most important operation in strawberry production and inefficiency during this very busy time can spoil the whole year's work. It has already been stressed that it is most unwise to grow more acreage than can be picked. The fruit becomes rapidly over-ripe and once the picking gets behindhand, it is almost impossible to continue, since rotting and over-ripe berries slow down the picking of the remainder of the crop.

Attempts are being made to design mechanical harvesting machines and evolve suitable varieties for them. It would appear that fruit needed for jam making may well be mechanically harvested within the next few years. However, varieties required for the fresh fruit market must be unbruised and as the prototype machines at present under test do result in some damage—as well as considerable wastage—it is likely to be many years before hand pickers become redundant.

ORGANIZATION IN THE FIELD

Female casual labour is necessarily employed for strawberry picking and often the only labour available is accompanied by babies, or toddlers below school age. These children can damage the crop and slow down the rate of picking. Where numbers suffice, growers have found it necessary to provide one woman to attend to the children, with a shelter, and a safe play area. In order to get enough pickers, growers may have to employ pick-up buses or run their own transport to collect the workers. Care should be taken that insurance is adequate in both cases.

School children may also be employed to pick fruit and, if properly supervised, can be useful when insufficient women are available. Their conditions and hours of work are strictly controlled by County byelaws and a copy of the regulations should be consulted.

The organization of the pickers depends on the arrangement of the field, but at all times the aim should be to keep the fruit collection point as central as possible. Care should be taken to organize the flow of empty trays and to remove full trays into a suitable shady place, or a cool store at regular intervals.

There is always a tendency for pickers, particularly those on piece work, to pick only the large berries, and supervision and checking of work in the field, as well as at the collection point, is essential.

Piece workers are often paid in cash at the collection point for each full, and satisfactorily picked, tray brought up. Sometimes a tally or token is given instead, redeemable in cash at the end of the day or week. Piece rates are based on picking rates per hour of 15 to 20 lb of ungraded fruit picked on the stalk, with higher prices paid for graded fruit.

Many growers prefer to work on a day-rate system with a bonus for each tray properly picked and presented. This method may well be the only way to get fruit carefully picked and graded, but it involves some ingenuity to arrive at a system that encourages rapid work at reasonable rates.

HARVESTING DESSERT FRUIT

The great majority of growers pick the fruit directly into ½ or 1 lb punnets already arranged in the market tray. This means that little or no grading of berries can be done; the pickers generally only leave very small, damaged, or rotted berries on the plant, or, if picked, drop this fruit in the alleyways. Although this method results in the maximum output per picker, it has the disadvantages that the tray may get dirty or wet, pickers tend to do a final 'top-up' of the punnets, using the bigger and more attractive berries, and often put damaged fruit into the punnets. If pickers are being paid piece work, i.e., so much per pound of fruit picked, or per completed tray, there is always a tendency to skimp the grading, handle the berries instead of the stalk and to pack them into punnets as quickly as possible. The only quality control that can be exercised under these conditions is at the collection point, where the trays should be examined as they are brought in by each picker. At the beginning and occasionally thereafter, in addition to a surface inspection, a punnet should be removed and the contents examined for bruising, retention of damaged fruit and uneven grading.

It is essential to exercise as strict a control as possible over the pickers, and those persistently bruising berries, or packing unsuitable berries, should be dismissed.

To some extent the crop sets the standard of picking and packing, since well grown, not too leafy plants, with readily seen, large berries are obviously easier to pick and grade than weedy, crowded plants with small or diseased berries.

The berry should never be handled or the surface will ultimately show bruises. Only the thumb and forefinger should be used to nip the stalk about ½ in. from the fruit and to hold the berry and lay it gently in the punnet. The biggest problem is to stop the pickers holding the berries.

Laying the fruit into the punnet requires experience. The berries are laid in the corners first and then when the bottom has been covered, the next layer is placed in position. The berries should be laid in with the stalk slightly downwards, so that the final appearance is one of the attractive colour of the berries with the minimum of green stalk and calyx showing. Some pickers are inclined to pack the top layer of berries stalk downwards. Although a most attractive appearance results, every berry has to be handled to be placed in position and must therefore be bruised.

Grading for the Dessert Market

Growers producing out of season and quality crops, generally endeavour to grade the fruit. Some form of grading may also be attempted by growers of main crop strawberries, who are concerned to pack to a better standard than the average, by placing all the small berries into one or two punnets in each tray, which can then be separately packed.

Grading may be done in the field, or in a packing shed. The latter is uncommon but is practised by some specialist producers. Each picker is provided with carriers which are generally wooden trays fitted with a handle. In the bottom of each tray is placed a folded cloth. The pickers are instructed to pick every berry that is two-thirds or more coloured even if it is damaged and to place it carefully in the tray in a single layer. As the trays are filled they are stacked and taken to the packing shed. This need not be elaborate but should be well shaded, dust free and airy. In the shed, tables are needed at a suitable working height. The more deft and experienced pickers are kept in the shed to repack the fruit. They have the market trays complete with punnets in front of them and rapidly transfer the berries to the appropriate grade, checking each fruit for damage and downgrading as necessary.

Although this method involves double handling of each berry it can ensure excellent standardized grades and all the damaged fruits are removed from the field. It is particularly valuable for specialist producers where labour is short and coming and going at all hours, as it enables those in the field to concentrate only on removal of fruit and not to need instruction in, and supervision of, grading. Where punnets are sold by weight—and many salesmen prefer this and may be able to obtain a price premium for guaranteed weighed punnets—this method has much to commend it. Slightly damaged fruit, with dry, slug or mechanical damage, may be graded out and sold as second grade; this fruit often meets a ready local market at a time of high prices when fruit is in short supply.

There are no statutory grades for strawberries*. Generally the largest berries, perfect in shape and completely unblemished, are put into an extra selected grade. It is usual to try to keep to at least a minimum of about 20–40 berries per pound for this grade, i.e., not more than 20 berries, and preferably less, per ½ pound punnet.

At the other end of the scale berries averaging more than 64 per pound are considered very small and more suitable for a jam grade. An admixture of large and small berries tends to look worse than a graded pack of medium sized berries.

As the strawberry is so perishable each berry should be examined for soundness; even the smallest infection by *Botrytis* will result in a wholly rotted berry by the morning and other berries in the punnet may become infected. Berries marked or damaged by boxes, or feet, slugs, beetles or birds must also be ruthlessly downgraded and discarded, or put in a separate punnet and not mixed with sound fruit. This is particularly important during wet weather when rots develop rapidly.

Fruit should not be left to get too ripe but picked when just fully pink. Varieties vary widely in the speed with which they become over-ripe. Some of the popularity of Cambridge Favourite and Redgauntlet is due to the fact that they remain firm ripe for several days without becoming too dark.

---

* Recommended European standard grades for strawberries have been prepared by the United Nations Economic Commission for Europe (AGRI/WP/1/EUR. STAN. 11) and copies are obtainable from H.M.S.O.

Cambridge Vigour rapidly becomes soft if slightly over-ripe and Cambridge Rival, although remaining extremely firm, becomes a very dark colour.

Ideally, the strawberry fields should be picked over daily in really hot weather, and two to three times weekly under cooler conditions. In practice, picking is often done only once per week, but as mentioned before, it is most unwise to grow more strawberries than there are available pickers. At least 10 to 15 pickers would be needed per acre for picking and grading an average crop into punnets; more workers would be required if more careful grading and packing was to be done. Picking rates into small punnets and grading the fruit average about 10 to 15 pounds per hour, but there can be great individual variation.

Tray Carriers

Also see Short Term Leaflet 81

Recently attention has been paid to the provision of carriers to hold the market trays, but it is still most usual to see the wooden, or fibreboard, market trays being used in the field. These trays are often rather wide for the rows and are difficult for the pickers to move, and they may get muddy or damaged.

The production of a light wirework or wooden carrier to hold these trays is strongly recommended (Plate VIII, bottom left). Studies by the N.A.A.S. in co-operation with Kent strawberry growers showed that a carrier high enough to go over the plant row rather than in the alley was the most suitable. The diagrams show the suggested size and construction recommended. The carriers may be made in wood, iron or aluminium alloy, but must be made as light as possible and strong enough to withstand the natural tendency of pickers to press on the handle when moving forward, or standing up. It is necessary to standardize the size of the market tray to be used in the carrier. Alternatively, the carrier can be made to hold a light wooden tray and the punnets be repacked into the market trays at the collection point from the picking trays. This permits some rough grading of fruit by the overall quality in each punnet but has the disadvantage that each punnet is handled twice. However, where punnets are to be sold by weight this method fits in well. A carrier holding a total of about 7 to 11 pounds of fruit in the appropriate $\frac{1}{2}$ or 1 lb punnets is quite big enough and not too heavy to move.

The carrier should be wide and high enough so that the legs do not damage the berries. Modifications could include narrow carriers for use between cloches and uneven legs for carriers to be used on very sloping sites.

If a carrier on legs is not adopted, wire, clip-on handles can be provided for wooden tomato trays where these are the market pack, and used to move the tray up the alleyway. Similarly, a wire framework with folding handle can be made to hold fibreboard market trays. This helps the picker move the tray and protects it from the ground.

The punnets should be placed into the market trays ahead of the pickers. Where there is plenty of storage space the market trays can be prepared before the picking season, or during wet weather, and stacked ready for the pickers. Even if this is not done it is best to have one woman preparing the trays—a useful job for an incompetent picker, or an older woman—rather than let all the pickers do this.

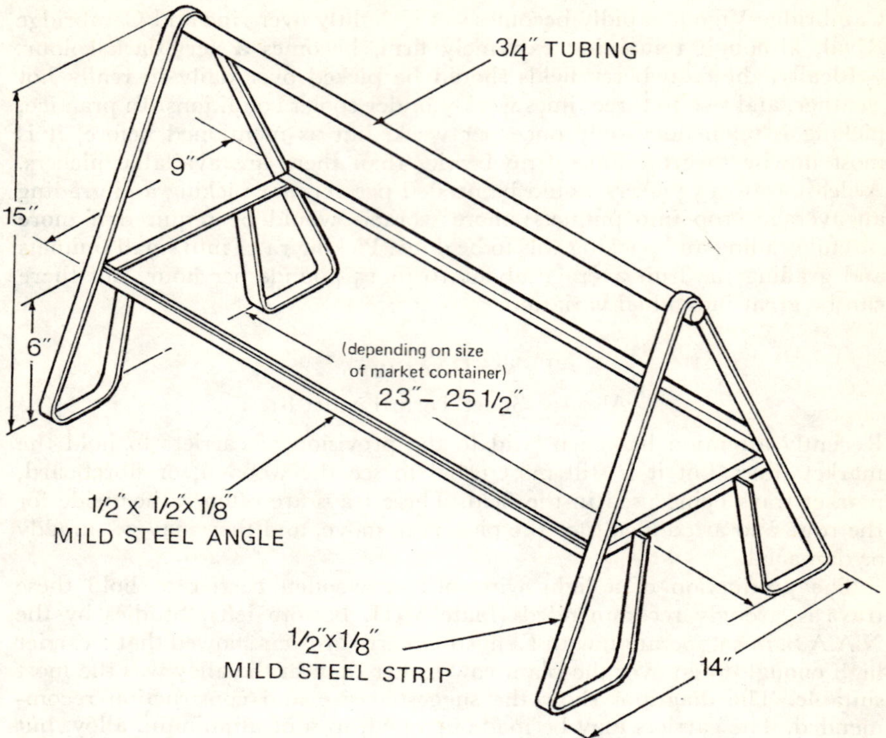

Fig. 3. Tray carrier for use when picking direct into punnets in market trays. Net weight of tray carrier: $4\frac{1}{2}$ lb.

Picking Trolleys

Picking trolleys enable the picker to be seated and to carry more empty and full trays. A few growers have made trolleys to their own design and several have been reported in use in Italy and Holland. The trolley has three wheels, set to the appropriate row width, two in one alley, one in the adjacent alley and a low seat for the picker who normally works backwards up the row, propelling the trolley by the feet and stooping to pick the fruit and place it in the punnet carrier conveniently positioned in front. Empty boxes can be carried behind the picker. Obviously the trolley cannot be used for work on uneven or stony ground. Growers thinking of making and using some trolleys should contact the N.A.A.S. for information.

HARVESTING FOR PROCESSING

Eight to ten pickers per acre is generally regarded as the minimum required in areas around Wisbech where the crop is sold almost entirely for processing. This labour is often referred to as van labour, being controlled by a gangmaster who is responsible for assembling the gang and through whom all business is conducted. Other pickers are obtained locally, or migratory labour is obtained from organized camps, e.g., student camps. In many cases the grower may have to provide or arrange transport in order to obtain workers.

Pickers are normally paid piece rates and the fruit is generally plugged in the field as picked. Generally, the foreman organizes the supply of trays into which the fruit is poured when weighing at the collection point, and pickers are then paid in cash or with token discs which are cashed at the end of the day.

Control of large gangs can be difficult and must be properly supervised so that all fruit is picked; this often necessitates a regular member of the farm staff being responsible for setting the pickers in the rows and controlling them as they move over the crop. Fruit may be either picked as plugged on the plant, or picked on stalk with the calyx attached. Higher contract prices are offered for plugged berries, and Cambridge Favourite, the main processing variety, plugs very easily on the plant. It is usual to pay pickers about $\frac{1}{2}$d ($\frac{5}{24}$p) per pound more for the plugged fruit. Varieties hard to plug, e.g., Huxley, are usually plugged later in the processing factory as a separate operation after picking.

The picking containers are often 4 lb baskets or trays provided by the grower. After picking, the baskets are carried by the picker to the weighing shelter where the fruit is poured into the processors' trays and is checked for rots and foreign matter. It is weighed and a payment made. The wooden or metal trays supplied by the processor are lined with greaseproof paper before being filled and hold 12 lb of fruit. These trays should be stacked either in the shade, or preferably collected at mid-day and again in the late afternoon and put into a cool building. Collection of picked fruit by the factory is normally made during the afternoon and evening. No grading is done for processed crops, except to throw out rots, damaged and over-ripe fruit. Matted rows, which is the normal method of growing strawberries for processing, are picked over continually by the gang of pickers working round and round the fruiting beds. The plants should be picked at least once a week, more frequently in hot weather, or in seasons when fruit is liable to rot.

Marketing for Fresh Consumption

THERE are three main channels of distribution to the consumer of fresh fruit:

1. through wholesale channels;
2. direct to shops and supermarkets;
3. direct to the public.

The bulk of strawberries for retail sale are consigned to wholesalers in the main population areas for sale on commission, but because of the perishable nature of the produce, direct sales to local shops, local wholesalers or to the public can be beneficial to the grower and provide the consumer with a fresher product.

Strawberries for the fresh fruit market are usually picked and packed direct into non-returnable containers called punnets or chips, which are transported to the markets in crates, trays or carriers. These outer containers may be returnable, or more usually non-returnable.

The costs of selling through wholesale markets will include the salesman's commission, usually 10 per cent of the gross selling price, and a market toll charge per packet, which can vary according to the size of the container and is about three to six pence (1¼ to 2½p). A porterage charge is frequently made even if the produce is collected by the purchaser. The grower also has to pay transport costs to the market, or where the wholesale salesmen arrange collection, there may be a collection charge. Invoices from the salesmen should show clearly the gross selling prices of the various packages of fruit and the charges which have been deducted. In some instances the salesman will provide returnable trays and covers, possibly stamped with his own brand name, the cost of which is sometimes included in the commission, or for negotiated charges.

Salesmen tend to specialize in the products they handle, and it is important for the grower to keep in touch with the requirements of the market and to choose salesmen and market centres most suitable for his transport arrangements and the type of fruit produced. When possible, the grower should visit the salesman during the marketing of his crop to assess how his fruit has travelled and to compare grading, quality and pack in relation to other market consignments. Growers should also encourage their salesmen to notify market requirements to them, i.e., to suggest when fruit is best packed in ½ or 1 lb punnets, as well as to comment on grading and quality of fruit as received.

Although punnets are commonly referred to in terms of the weight of fruit they hold, there is no statutory requirement for a grower to sell by weight when selling through wholesale channels. It is in fact a common practice, which is accepted in the markets, for wholesale sales to be by the punnet—i.e., by volume—without the contents being weighed. However, if the fruit is sold by weight and marked accordingly, then the contents must be made up to the correct weight for the particular size of punnet, and must correspond with any marking of the weight. All retail sales (including direct sales by a grower to a retail customer) must be by weight, and the customer must be informed of the weight before he pays for, or takes possession of, the fruit. The way in which this may be done depends upon the weight of the empty punnet. Where this does not exceed the amount shown in the table, the fruit may be weighed in the punnet, and the gross weight marked on a ticket or on the punnet, or a special weighing machine may be made available for the customer's own use. Where the weight of the empty punnet exceeds the amount listed, the net weight of the contents must be marked on the punnet.

Gross weight of punnet and fruit (lb)	Weight of empty punnet (oz)
½	1
1	1¾
2	3½
3	4½
4	6
5	7½
6	9
Exceeding 6	1 oz per lb

Although direct selling to multiple stores or supermarkets has only been tried on a limited scale, many growers sell all, or a large part of, their crop direct to retail shops, or from roadside stalls. This direct trading takes up time, needs careful organization and may interfere with other work.

In some areas where labour is difficult to obtain, an increasing amount of fruit is being sold on a pick-your-own basis. The grower uses his own labour to pick the better fruit during the start of the season and when the size of berries is rather small, or if the prices are disappointing, he allows purchasers to pick their own fruit. This method needs careful supervision and organization. Containers (punnets or baskets) may be sold to intending purchasers at the point of entering—or their own containers weighed and a weight note given. The area to be picked should be marked off with posts and string, or the fruit will not all be cleared and plants may be damaged. Supervisors must be in constant attendance at the departure point to weigh the fruit and collect the money and also at the fruiting beds to supervise movement over the rows. Parking for cars may have to be provided and controlled to avoid contravening traffic regulations. It is usual to advertise this fruit picking in local papers, giving the times of availability and charges, which are generally at the prevailing retail price. With proper organization this method can be rewarding, particularly as fruit is most required over the weekends, when market prices are lowest, but it should be regarded as supplementary to organized marketing.

CONTAINERS

The non-returnable punnets or chips are made of white or printed fibreboard, waxed cardboard, wood pulp, or plastic. Very few wood veneer punnets are used today because of their cost. Handled baskets, made of waxboard or veneer usually holding 2 lb of fruit, are sometimes used, but more generally for direct or roadside trade only, since the risk of damage to the fruit is increased when these containers are sold through wholesale channels. The very earliest fruit is marketed in $\frac{1}{4}$ lb punnets. Quality fruit during the main season is sold in $\frac{1}{2}$ lb and 1 lb punnets. When the decimal system is introduced, the units of sale are expected to be 125g, 250g, and 500g. Some producers of high quality fruit cover each individual punnet with a transparent plastic cover, held with a rubber band.

Over the last few years non-returnable fibreboard punnet carriers and specially designed non-returnable wooden market trays have been used, but a large quantity of fruit is still sent in the Dutch type tomato tray, either new, or reclaimed after one journey with tomatoes. Very few returnable wooden crates are now in use. The Dutch type tomato tray will hold 10 or 11 $\frac{1}{2}$-lb oblong punnets or 15 $\frac{1}{4}$-lb punnets; it is not suitable for 1-lb punnets. The presence of corner pegs in this tray makes it difficult to fit in the punnets satisfactory, but the pegs help to provide good ventilation when the trays are stacked, and allow high stacking with greater safety.

The British Standards Institution has prepared a revised standard (BS: 3050: 1967) for punnets and containers which will fit conveniently on to the 40 × 48 in. international standard pallet that has been generally accepted for horticultural produce. This standard sets out a complete specification for wooden punnet containers and lays down performance requirements for punnet containers made of other materials. The recommended sizes for the containers are external base dimensions of 20 × 16 in. with an internal depth of $3\frac{1}{2}$ in. for $\frac{1}{4}$-lb and $\frac{1}{2}$-lb punnets, and of 4 in. for 1-lb punnets.

The market containers may be covered with paper or transparent covers, frequently provided by the market salesman. Transparent covers are a safeguard against pilfering and allow fruit to be seen and yet kept free of dust. The material used should be moisture permeable, e.g., cellulose acetate and the P.T. grade of cellulose film, or ventilation holes should be provided, otherwise the fruit may overheat. In some markets it is customary for two to three containers to be tied together and only the top one covered.

Number of punnets (gross) required per ton of fruit, to nearest gross:

$\frac{1}{4}$ lb = 63 gross
$\frac{1}{2}$ lb = 32 gross
1 lb = 16 gross

TRANSPORT TO MARKET

Road transport is most commonly used for short distances, or when the fruit can be taken direct to the markets or shops and intermediate handling avoided. For longer distances rail transport is more common, and special trains made up of ventilated bogie vans are provided to serve the main growing areas. For distribution of such perishable fruit, especially high-grade dessert produce, it is essential to have proper organization from grower to seller. Any unnecessary delays should be avoided, especially in hot or muggy weather when greatest wastage and deterioration of fruit can take place.

Once the fruit has been picked it should be quickly taken into the shade. Fruit packed ready for despatch should never be stacked unprotected in the field, but put where there is free air movement, out of direct sun and where the temperature is not higher than the air shade temperature in the open. Preferably the fruit should be moved regularly to a cool place, such as a fan ventilated room or refrigerated store, e.g., an apple or pear store, to lower the temperature by removing the field heat of the fruit.

TRANSPORT

The marketing of strawberries can be materially assisted with modern developments in the application of refrigeration. Removal of field heat from the fruit by putting it into a pre-cooling chamber shortly after picking, so reducing the temperature of the berries, and then transporting it in closed, insulated containers, slows down deterioration and allows the fruit to be transported long distances. This pre-cooling is commonly used in the U.S.A., the fruit being cooled to 1·7° to 4·4°C (35° to 40°F).* Mechanically refrigerated vehicles for the transport of fruit are not generally available in the United Kingdom, nor are they essential for the shorter distances normally covered here.

---

* More detailed information on pre-cooling is given in the leaflet *The travelling and pre-cooling, transport and storage of strawberries and raspberries* by W. Hugh Smith, obtainable free from Ditton Laboratory, East Malling Research Station, Maidstone, Kent.

Fruit that travels in insulated containers will 'sweat' (heavy condensation of moisture from the atmosphere over the fruit) on removal to a warmer atmosphere and it is advisable that it should have reached a temperature of 10°C (50°F) by the time it is unloaded. Under humid conditions, fruit, even a few degrees below air temperature, will acquire a dullness due to condensation.

Although the use of insulated and refrigerated containers is not usual in England, some use is made of dry ice (solid carbon dioxide). This is used mainly to build up the carbon dioxide, within a reasonably tightly sealed container, as this has been shown to prevent rotting of the fruit. The amount of dry ice used is not sufficient to lower the temperature of the fruit when loaded, so if possible the fruit should first be pre-cooled, but it will help prevent an increase in temperature. The dry ice is best placed within some empty trays, stacked at the top of the stack, and not too close to the surrounding fruit.

If no use is being made of refrigerated containers, or carbon dioxide atmosphere, then every care should be taken to keep the fruit as cool as practicable. Fruit should be stacked in the shade, and lorries or rail vans loaded to allow the maximum ventilation between the trays and also kept in the shade as far as practicable. Night temperatures during the strawberry picking season are often considerably lower than day temperatures and fruit transported during the night can be helped to cool. Fruit picked in the morning and evening will be at a lower field temperature than that picked during the day, and should be selected for the most distant markets. The type of tray chosen to hold the punnets should allow good aeration of the berries.

Strawberries cannot be stored. They may be held at the longest for one or two days in a temperature of 1·7° to 4·4°C (35° to 40°F), but even under these cool conditions they deteriorate rapidly on removal to normal air temperatures. However, the provision of a refrigerated store plus the use of carbon dioxide may enable a heavy pick from Sunday or Monday to be held for marketing on Wednesday and thus to avoid the low prices common on Monday and Tuesday when markets are often over supplied. The salesman should be notified if fruit has been stored.

FRUIT FOR PROCESSING

Fruit for jam making, canning or freezing is sold by the ton on a growing contract of three to four years, or on an annual contract made with the processing industry. This fruit is picked direct into the metal or wooden trays supplied by the merchant and collected for delivery to the factory by the processor, or merchants who act as buying agents.

Whatever form of disposal is adopted, good quality fruit presented in a clean and attractive manner will generally receive the best prices.

Appendix

TARIFFS (as at 1.1.1970)

IMPORTED fresh strawberries carry a tariff of £1 17s. 4d. (£1·86$\frac{2}{3}$) per cwt (4d. (1$\frac{2}{3}$p) per pound) from 1st to 9th June and £2 16s. od. (£2·80) per cwt (6d. (2$\frac{1}{2}$p) per pound) from 10th June to 31st July. For the remainder of the year the tariff is 10 per cent of the value (c.i.f.). Frozen unsweetened strawberries have a tariff of 15s. od. (75p) per cwt all the year and strawberries provisionally preserved, for example, with sulphur dioxide etc. are also charged 15s. od. (75p) cwt all the year. Commonwealth preference areas are free from all these duties.

SHORT TERM LEAFLETS

These S.T.L. are obtainable from advisers of the National Agricultural Advisory Service on request, or in conjunction with advisory enquiries. They are revised more frequently than this Bulletin and so can contain more up-to-date information. Leaflets dealing with strawberries are:

23 Chemical Weed Control in Strawberries
54 Cold Storage of Strawberry Runners and Their Use
64 Control of Botrytis Fruit Rot of Strawberry
77 Promising Cultivars of Hardy Fruit
78 Continuous Polythene Tunnels for Early Strawberry Production
81 Harvesting Strawberries: A Work Study Investigation in Kent

ADVISORY LEAFLETS

47 Wingless Weevils
226 Red Spider Mite
410 Red Core of Strawberries
414 Eelworms on Strawberries
530 Virus Disease of Strawberries
564 Soil-borne Virus Diseases of Fruit Plants

Single copies of Advisory Leaflets and Short Term Leaflets are also obtainable free of charge from the Ministry of Agriculture, Fisheries and Food (Publications), Government Buildings, Block C, Tolcarne Drive, Pinner, Middlesex, HA5 2DT.

BULLETINS

These are obtainable through any bookseller or direct from H.M.S.O., P.O. Box 569, London S.E.1.

Soils and Manures for Fruit (Bulletin 107) 6s. 6d. (32$\frac{1}{2}$p) (postage extra)
Fruit Spraying Machines (Bulletin 5) 6s. (30p) (postage extra)

Index

Printed in England for Her Majesty's Stationery Office, by Elliott Bros. & Yeoman Ltd., Liverpool
3753 Dd. 147081 5000 K 40 2/70 Gp 474

MINISTRY OF

AGRICULTURE, FISHERIES AND FOOD

Publications on Pests and Diseases

Common Names of British Insect and Other Pests (Technical Bulletin 6)	10s. (10s. 6d.)
Diseases and Pests on Horticultural Planting Material	12s. (12s. 8d.)
Sugar Beet Diseases (Bulletin 142)	7s. 6d. (8s.)
Diseases of Bees (Bulletin 100) ..	7s. (7s. 4d.)
Identification of Common Water Weeds (Bulletin 183)	4s. 9d. (5s. 3d.)
Diseases of Bulbs (Bulletin 117) ..	5s. (5s. 8d.)
Mites of Stored Food (Technical Bulletin 9)	17s. 6d. (18s. 8d.)

Prices in brackets include postage

Government publications can be bought from the Government Bookshops in London (post orders to P.O. Box 569, S.E.1) Edinburgh, Cardiff, Belfast, Manchester, Birmingham and Bristol, or through any bookseller